Gone Fishin'

The
100 Best Spots
in New Jersey

Manny Luftglass and Ron Bern

Rutgers University Press
New Brunswick, New Jersey

Second paperback printing, 2003

Library of Congress Cataloging-in-Publication Data

Luftglass, Manny, 1935–
    Gone fishin' : the 100 best spots in New Jersey / Manny Luftglass and
Ron Bern.
      p.   cm.
    Includes bibliographical references and index.
    ISBN 0-8135-2595-0 (pbk. : alk. paper)
    1. Fishing—New Jersey. I. Bern, Ronald Lawrence, 1936– . II. Title.
SH525.L84   1999
799.1'09749—dc21   98-22950
                                         CIP

British Cataloging-in-Publication data for this book is available
from the British Library

Manufactured in the United States of America

Gone Fishin'

Contents

Illustrations

Foreword

Fishing has never been better in New Jersey. At no time in our state's history has there been a greater diversity and abundance of fish and fishing opportunities. We can attribute this to many factors. Our water quality has improved significantly since the passage of the Clean Water Act and the achievements of state programs for water resource protection, restoration, and enforcement. New Jersey's intensive fish rearing, stocking, and management of our marine and freshwater fish, and our initiatives for increasing fishing access, creating new reservoirs, promoting catch and release, and the support from our fishing organization partners have had a major positive impact. All of this combined with some long-term help from Mother Nature has made our state a great fishing destination.

New Jersey has an astounding quantity and diversity of rivers, streams, lakes, ponds, and 120 miles of ocean coastline, and over 390,000 acres of estuarine area and inlets spread all along the coast. We have over 4,000 reservoirs, lakes, and ponds larger than one acre that provide over 61,000 acres of open water. In addition to bountiful open water, there are over 6,400 miles of rivers and streams including the outstanding Delaware River, which is rapidly becoming one of the top fishing spots in the east. While we may be the fourth smallest state in the nation, we are in a unique geographic position. Our state lies right where the northern ecotypes reach their southern range limit and the southern ecotypes reach their northern range limit. This location provides us with an incredible variety of water from the crystal clear glacial lakes and rocky streams in the north counties, to the fertile ponds and reservoirs of central Jersey, to the tea-stained acid waters in our southern counties, not to mention our vast coastal waters. Of the fourteen Atlantic coastal states, only Florida and North Carolina can boast of a larger recreational marine fishery. There's water for every type of fish and every type of angler.

To make the most of nature's work in our state, over the years we have worked to open over four hundred of our freshwater bodies covering about 25,000 acres and hundreds of miles of rivers, streams, coastline, and bays

Figure 1 ■ New Jersey Fish and Wildlife Director Martin McHugh and his son Marty teamed up to catch this nice rainbow trout in the Pequest Fishing Education Pond on the opening day of trout season.
(Photo: N.J. Division of Fish and Wildlife)

to public angling. Public access has increased dramatically in the recent past and there are efforts to increase this even further in the near future thanks to the support of Governor McGreevey, Department of Environmental Protection Commissioner Campbell, and the DEP's Green Acres Program, which is working with our Division of Fish and Wildlife. Obviously, an important aspect of increased fishing access includes our public boat launches program. Currently we have hundreds of public ramps and we are striving for more. We've purchased private ramps, built ramps on public land, and are actively working with municipalities to develop and operate boat ramps throughout the state.

In addition to the work being done to restore our freshwater habitats, an intensive program of artificial reef construction, started in 1984, has created a network of artificial reefs in the ocean waters along the New Jersey coast that provides productive habitat for fish, shellfish, and crustaceans, fishing grounds for anglers, and underwater structures for scuba divers. The reefs now cover more than 25 square miles of sea floor and are strategically located along the coast so that at least one site is within easy boat range for our off-shore anglers out of twelve New Jersey ocean inlets.

If the size of fish is a barometer of fishery conditions and fishing opportunity, there are many good reasons to be proud and excited about fishing in New Jersey. Our anglers have caught thirty-seven state record fish in the last ten years. Fisheries for five of our state record freshwater fish did not even exist in New Jersey fifteen years ago. Our Freshwater Fisheries Program has recently established viable New Jersey fisheries for muskellunge, walleyes, hybrid striped bass, lake trout, and northern pike. Many of those state records are impressive fish for any state in the country. Recent records include the 42 lb. 13 oz. muskellunge caught in 1997, a 32 lb. 8 oz. lake trout in 2002, a 13 lb. 9 oz. walleye in 1993, a 27 lb. bluefish in 1997, a 87 lb. cobia in 1999, a 105 lb. black drum in 1995, and a 25 lb. tautog caught in 1998 that is a world record fish. New Jersey saltwater anglers hold four other world records including a 78 lb. 8 oz. striped bass caught in 1982.

The freshwater stocking program annually places over 3 million fish in New Jersey's waterways. The state-of-the-art Pequest Trout Hatchery each year produces over 700,000 trout that are stocked in the spring, fall, and winter. The number, size, and health of the trout stocked today have never been better. In addition to the trout raised at Pequest, we have developed and expanded both warmwater and coolwater fish production at the Hackettstown Hatchery which now annually raises and stocks 1.5 million bass, bluegill, catfish, crappie, lake trout, muskellunge, pike, and walleye. The new facilities at the Hackettstown Hatchery allow us to raise fish that are healthier and up to 37 percent larger in the same amount of time it took using the old facilities. Stocking larger fish means more fish in the future for Garden State anglers because the size at stocking is a key factor in how many fish will survive and mature into adults.

The improvement of our water quality and the development and nurturing of our fantastic fisheries is a great New Jersey success story. Much of that success is due to the support and efforts of the many anglers and conservationists that belong to organizations such as BASS, Jersey Coast Anglers, Knee Deep Club, Muskies Inc., Paradise Fishing Club, Recreational Fishing Alliance, Round Valley Trout Association, Trout Unlimited, and numerous local groups. They are on the front line in the fight to preserve and enhance our aquatic resources and have donated thousands of hours and dollars to make fishing better in our state.

For a small and highly populated state we are truly fortunate to have the aquatic diversity and abundance of fishable water that we have in New Jersey. Also, for a small state our fishery is incredibly diverse and improving in quality each year. Take some time with a friend or a family member to explore the fishing possibilities New Jersey has to offer. I am confident you will be pleased with what you find and catch!

Martin McHugh
Director
Division of Fish and Wildlife

Authors' Preface

New Jersey is our home, not by accident of birth or commercial circumstance but because this is where we consciously chose to live and raise our families. Like so many of our neighbors, we are not natives. Manny Luftglass was born and raised in Brooklyn; Ron Bern in the cotton country of South Carolina. We migrated to central Jersey in 1964 and have never regretted making this our home. Now that our children are grown, we choose to remain here because of all this state has to offer.

People from other parts of the country tend to think of New Jersey in terms of smokestacks and crowded highways and tired urban communities. However, we know the sheer beauty and the bounty that is New Jersey outdoors. This book is written to share what we have learned in more than three decades of fishing here—sweet water, brackish, and salt.

We have explored only 100 spots, and we acknowledge (1) this only scratches the surface and (2) we may have missed your favorite spot. However, our goal was to make this a representative sample, including large rivers and small streams, massive reservoirs and small ponds, surf-casting spots and saltwater bays, ocean spots and artificial reefs. We think we got it about right. And as a little bonus, we've added "secret tips" that can help you catch more fish.

We have arranged each category of fishing spots on a more or less north-to-south, west-to-east axis for your convenience. We have not gone into detail about fishing or boating regulations primarily because the "regs" are subject to change. At any rate, you will get a detailed compendium of regulations when you purchase your freshwater fishing license. A separate booklet of saltwater rules also is available free of charge at most tackle stores.

We generally fish together; just the two of us. But this time we would like you to join us—beginning quite appropriately on the Delaware River, where our fishing partnership began a third of a century ago.

Gone Fishin'

Key

1 Greenwood Lake
2 Shepard Lake
3 Upper Wanaque River
4 Monksville Reservoir
5 Big Flatbrook
6 Canistear Reservoir
7 Clinton Reservoir
8 Swartswood Lakes
9 Paulinskill River
10 Ramapo River
11 Hudson River
12 Drake Lake
13 Wawayanda Lake
14 Pompton Lake
15 Halendon Reservoir
16 Lake Hopatcong
17 Pequest River
18 Rockaway River
19 Passaic River
20 Musconetcong River
21 Budd Lake
22 Black River
23 Raritan River, North Branch
24 Raritan River, South Branch
25 Merrill Creek Reservoir
26 Spruce Run Reservoir
27 Rahway River
28 Round Valley Reservoir
29 Delaware River
30 Raritan River, Main Body
31 Millstone River
32 North Beach
33 Raritan Bay
34 South River
35 Cheesequake/Hooks Creek Lake
36 Sandy Hook Bay
37 Seventeen Fathoms
38 Scotland Buoy
39 Farrington Lake
40 Lefferts Lake
41 B.A. Buoy
42 Amwell Lake
43 Navesink River
44 Mud Buoy
45 Sandy Hook Artificial Reef
46 Delaware and Raritan Canal
47 Carnegie Lake
48 Molder's Ponds
49 Shrewsbury River
50 Shrewsbury Rocks

51 Marine Place
52 Lake Mercer
53 The Farms
54 Assunpink Lake
55 Rising Sun Lake
56 Stone Tavern Lake
57 Texas Tower
58 Shark River
59 Klondike Banks
60 Manasquan Reservoir
61 Sea Girt Artificial Reef
62 Metedeconk River, North Branch
63 Manasquan River
64 Manasquan Inlet
65 Point Pleasant
66 Manasquan Ridge
67 Hudson Canyon
68 Normandy Beach
69 Rancocas Creek
70 Mirror Lake
71 Seaside Heights
72 Toms River
73 Forked River/Oyster Creek
74 Island Beach State Park
75 D.O.D. Lake
76 Barnegat Bay/Long Beach Island
77 Barnegat Inlet
78 Garden State North Artificial Reef
79 Barnegat Ridge
80 Loveladies/Coast Avenue
81 Atlantic City Area
82 Iona Lake
83 Garden State South Artificial Reef
84 Mullica River
85 Great Bay
86 Parvin Lake
87 Lenape Lake
88 Union Lake
89 Harrisonville Lake
90 Atlantic City Surf and Rocks
91 Atlantic City Artificial Reef
92 Cohansey River
93 Fortesque
94 Maurice River
95 Bauer's Fishing Preserve
96 Sea Isle City
97 Delaware Bay
98 Cape May
99 Cape May Rips
100 Cape May Artificial Reef

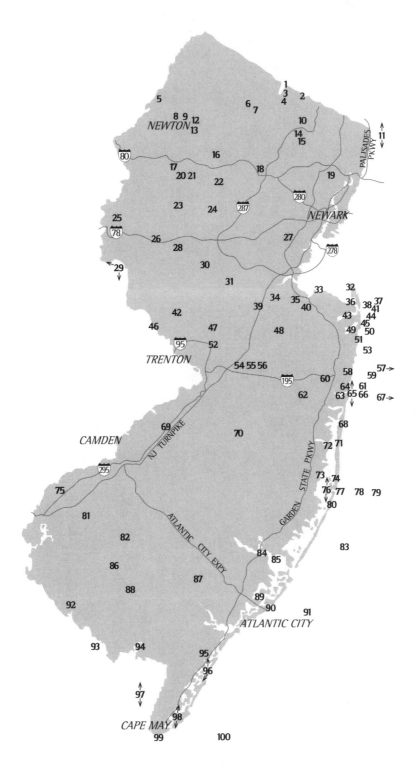

Freshwater
Fishing

Major .... *Rivers*

29 Delaware River

The Delaware River is a category unto itself, providing fishing that rivals any body of moving water in America. The river regularly yields state-record fish (walleye, inland striped bass, American shad, and tiger muskellunge at this writing). It shelters superb populations of smallmouth bass, largemouth bass, true muskellunge, white perch, catfish, sunfish, walleye, trout, and huge carp, plus spectacular runs of shad and herring.

The noted fishing writer Ed Zern once wrote that if forced to choose one body of fresh water to fish for the rest of his life, it would be the Delaware River. Noting that he had fished the finest rivers, lakes, and streams in countries throughout the world, he said that none equaled the Delaware in its extraordinary variety of fishing opportunities, from its sparkling fresh headwaters to the salt waters of Delaware Bay. We could hardly agree more wholeheartedly. Indeed, the long-standing fishing partnership of the authors was forged along the middle stretches of the Delaware, where we fished the deepest channels for massive carp, smallmouth bass, sleek channel catfish, fat bluegills by the hundreds, and even the occasional striped bass. From its headwaters to the sea, the Delaware runs through four identifiable sections, each with its own character and

Figure 2 ■ Dave Au with 30.12-pound Delaware River freshwater striped bass.
(Photo: N.J. Division of Fish, Game, and Wildlife.)

superb angling possibilities. The first three are examined here; the fourth—Delaware Bay—can be found in our "Saltwater Rivers and Bays" chapter. By examining the "Big D" in sections, we help to unravel its changing personality and grand angling opportunities.

■ Delaware River, Section One

Directions: Route 97 near Callicoon, N.Y., and at Narrowsburg, N.Y. Access for fishing and boat launching. Also Route 371 at Damascus, Pa., right at the bridge. Also at Depew, N.J., off Old Mine Road, approximately 9.3 miles north of the Delaware Water Gap. Also the Eshback Access Site approximately 6 miles south of Dingman's Ferry Bridge off Pa. Route 209.

The headwaters of the river begin at the junction of the river's east and west branches, both beginning in high mountains of New York State. The river's west branch begins at Pepacton Reservoir and its east branch at the Cannonsville Reservoir. The outflows from these two large reservoirs are themselves home to huge populations of native brown and rainbow trout.

The branches join to form the Delaware at Buckingham, Pennsylvania, on its western side and Hancock, New York, on its eastern side, thereupon forming a part of the boundary between the two states. The upper section, comprising eighty miles of wild, beautiful river, is especially known for trout fishing and regularly produces trophy trout. This excellent fishery enjoys special environmental conditions: especially controlled releases of cold, clean water from the Pepacton and Cannonsville Reservoirs that keep water temperatures low during warm weather.

Shad in considerable numbers also run all the way through to the upper reaches of the river. The shad run here generally lasts much later than runs downriver because of the colder water. The area referred to as the "Eshback Access," located six miles south of Dingman's Ferry, is a fabled shad hot spot, especially along Walpack Bend and in the relatively deep mile-long stretch just upriver from the access point. Also due to its abundance of cold, clean water, the upper section of the river is well known for its smallmouth bass and walleye fishing.

Successful trout anglers in the upper reaches of the Delaware include many purists casting flies that match an abundant hatch of marine insects (little black caddis, red quills, and black stoneflies emerge as early as mid-April; green midges, white caddis, and deerflies are still emerging in July). However, a healthy share of trout are taken by spin casters fishing baby night crawlers, salmon eggs, and corn kernels along promising stretches of bottom.

Smallmouth bass are taken virtually throughout the year, including the warmest months of summer, especially around points and drop-offs. Like trout, these aggressive, hard-fighting fish have a wide range of appetites. Their favorite foods are hellgrammites, small crayfish, and fallfish (baby "chub"). However, they are also taken on night crawlers and a wide range of artificials (small Rapalas and silver Mepps are our personal favorites), especially in rocky sections of the river.

Walleyes are less familiar to anglers in the river but offer great sport to those willing to persevere in often-difficult conditions. These begin with weather. The axiom favoring walleye fishing recalls duck hunting; that is, the worse the weather, the better the hunt. Walleyes tend to be caught in cold, blustery weather, with the season beginning in February and virtually ending with the opening of trout season in April and starting again when waters cool in late fall. Many walleyes are "accidentally" taken on shad darts, but most are taken on live bait, including night crawlers, minnows, and leeches. The state-record walleye, weighing 13 pounds 9 ounces, was caught in the Delaware by George Fundell, Jr., in 1993, no doubt in uncomfortable weather.

Tip: Brave the elements to fish for walleye right through a dark, cold winter's night.

Muskellunge also are taken in the upper reaches of the river, most often on heavy gear and a wide range of artificials. However, the muskie fisherman who ignores live suckers and exceptionally large fallfish may be missing a good bet. The stocking of muskellunge in the upper reaches of the river began as an experiment more than twenty years ago. Today the occasional lunging strike and spectacular fight of a big muskie is making this another success story for the New Jersey Division of Fish, Game, and Wildlife. To underscore this success, the former New Jersey state-record muskie, tipping the scales at 38 pounds 4 ounces, was taken in this section of the river in 1990. (This record stood until 1997, when Bob Neals landed a 42-pound 13-ounce muskellunge at Monksville Reservoir.)

All in all, the river's first section is almost unexcelled as a sportfishing venue. It is certainly part of what Ed Zern had in mind when he praised the Delaware so unstintingly for its angling variety and excitement.

■ Delaware River, Section Two

Directions: Access at Delaware Water Gap off Route I-80 at the Kittatinny Visitor Center. Bull's Island access is off Route 29 approximately 3 miles north of the Stockton, N.J./Pa. Route 263 bridge. Belvidere launch reached via exit from Route 46, Martins Creek, Pa., via secondary paved roads.

A north-south split occurs in the river at the Delaware Water Gap, a natural wonder in the wild and beautiful Kittatinny Mountains in the northwest corner of New Jersey. The Water Gap was created by the sculpting power of the river over the course of 150 million years. It now flows clean, deep, and dark through a steep, rock-faced ravine some 1,600 feet deep.

More than 120 years ago, a writer in the *New York Times* noted: "Nature presents herself here in so many phases of gentle beauty and wild savagery that even scant justice would hardly be possible in attempting a description of the scenery of the Gap."

Here begins a fifty-mile stretch of exceptionally clean and bountiful water, which is notable for the varieties of its fish, especially large carp. In one particular mile-long span of the river in the vicinity of Bull's Island State Park, the authors have caught and released no fewer than 1,250 carp, many of them in double figures and the four largest weighing 15 pounds, 17 pounds, 17 pounds 4 ounces, and 20+ pounds. (We apologize for the inexactitude of the final weight; our freshwater fishing scale registers weights only up to 18 pounds and this largest fish jerked it through its highest register.) We recall one particular July Fourth morning when we landed 22 carp fishing from shore at Byram, not one weighing less than 5 pounds.

The largest of the fish described above illustrates our fishing partnership in all regards. We compete hotly for most fish and largest fish; indeed, it is our competition that provides a context for understanding our fishing partnership. Yet either of us will do virtually anything to help the other land a fish. For example, RB ran into trouble fighting this massive carp, fishing it as he was from shore under a thicket of tree limbs, which made it impossible to get his rod tip high enough to fight the fish effectively. Sizing up the situation, ML instantly began climbing the offending tree and, as RB fought the fish from his knees, began to tear off limbs with his bare hands. He cleared a sufficient space and by his good offices, the fish was landed, weighed, and released.

Tip: When fishing for carp, here and in other moving water, double-anchor your boat across the current and cast straight downstream, using light to medium spinning tackle and 6- to 8-pound test line. The terminal rig is a #6 Eagle Claw style baitholder hook, tied 12 to 18 inches below a barrel sinker (usually a half to three-quarters of an ounce) which ranges freely above a small split shot. Bait is either cooked and flavored (cornmeal) carp bait or four to six kernels of canned corn. It is useful to chum upstream in the current, especially with bits of crumbled cornmeal bait, and to keep your drag wide open in anticipation of the strike.

Carp are by no means the only superb fish in these waters, although they are on average the largest and strongest. Indeed, tiger muskies have been extensively stocked in the Frenchtown/Bull's Island/Lambertville stretch of the river. (The state-record tiger muskie, tipping the scales at 29 pounds, was taken in the Delaware by Larry Migliarese in 1990.)

This stretch of the river is also prime shad and herring water, especially from Lambertville upriver to the wide bend at Byram. Moreover, the deep channel across from Bull's Island Park, which reaches depths of 18 feet in spots, attracts feisty smallmouth bass in numbers, especially in the summer and early fall. In addition, anglers here catch three varieties of catfish (including large and powerful channel cats), white perch, four varieties of sunfish, and an occasional striped bass in these waters, along with fallfish, suckers, and surprisingly powerful American eels.

Boat-launch facilities are available, among others, north of the mouth of the Delaware and Raritan Canal at Raven Rock, off Route 29 approximately 1 mile north of the entrance to Bull's Island State Park, right at our favorite stretch of our favorite river in New Jersey.

■ Delaware River, Section Three

Directions: Access at Washington Crossing State Park, directly off the north side of Route 546, 0.5 miles west of Route 579. Some shore access via N.J. Route 29.

The beginning of the third section of the river is marked by two towns with identical names, facing each other across the river. The towns are, respectively, Washington Crossing, Pennsylvania, and Washington Crossing, New Jersey. The state park at Washington Crossing is a promising spot to begin fishing this stretch. Access is good here and fishing is often excellent. Several years ago, in fact, an angler named Larry Migliarese landed the 29-pound state-record tiger muskie from shore right on park grounds.

Brackish waters begin in this general area. The predominant fishing in this section of the Delaware features stripers, largemouth, crappie, and muskies, with fine seasonal action on shad, herring, and carp. All fishermen along the Delaware owe a great vote of thanks to determined shad fishermen who fought long and hard to prevent oxygen-depleting pollution from spoiling this great fishery. In the seventy years beginning just before the turn of the century, the annual catch of shad in the Delaware, from the bay and tidal river up to the headwaters, had dropped from 15 million pounds to less than a 100,000 pounds. Alarmed sportsmen's organizations joined environmental groups in demanding—and eventually forcing—a halt to chemical pollution. Today, the Delaware is a vastly cleaner, healthier river, with massive runs of shad and herring powering upriver to spawn.

The shad run is brief, sometimes lasting no more than a few weeks in April and May. When the shad are running, hundreds of boats anchor "shoulder to shoulder" in the channels and along the drop-offs in quest of big roes and bucks.

Tip: When fishing for shad, single-anchor in fast-moving water. Use down-riggers to hold flutter spoons just off bottom for best results.

At about the same time, herring appear in the river in huge numbers and are commonly caught on "trees" of bare gold hooks.

Food sources are the primary reason for the presence of exceptional bass and muskie fishing in this stretch of the river. Shoals of herring and shad fry sparkle through the river's shallows throughout the summer and in early fall. Bunker, spot, and snapper blues—all major food sources for striped bass—work up the stretches of the lower tidal river. In winter, when the herring and shad fry run out into the ocean, largemouth bass become largely inactive in the colder water. However, stripers pick up some of the slack, hitting everything from plugs and bucktail jigs to live eels and live white perch in the creek mouths, around bridge pilings and in the drop-offs to deep holes. As in most forms of bass fishing, structure is an important key, with coves and pilings meriting special attention.

In the spring, when bass begin to spawn, tiger muskies patrol the coves in search of herring or larger fish too distracted by the business of procreation to note their presence. Spring is a perfect time to hook the big tiger muskie you've been looking for, from a boat or from shore. Big schools of stripers move up the river in spring to spawn as well, and continue to feed through summer right into early fall. When the water begins to cool late in the fall, the cycle begins again, with all three prime game fish in this stretch of the river feeding downstream on young shad and herring heading out to sea.

Largemouth bass, shad, and big, powerful carp also abound in these waters, and there are plenty of boat ramps, including the Burlington ramp next to Assiscunk Creek, the ramp in Westville on Big Timber Creek, and down by the Commodore Barry Bridge on Raccoon Creek.

11 Hudson River

Directions: Take the Palisades Parkway north from its inception, two miles north of the George Washington Bridge, and follow along the river to marked entry points and local roads.

The mighty Hudson River originates at Lake Tear of the Clouds, the highest point in New York's Adirondack Mountains, and steadily gathers force and character as it flows 315 miles to New York Harbor. One of the geological wonders of the world is the 40-mile-long Palisades on the New Jersey side of the Hudson River, a sheer rock wall as high as 550 feet, north of Fort Lee. Once the river was badly polluted. Today, however, the worst polluters have been challenged and stopped, and the river flows cleaner than it has for decades. There is a great abundance of fish in the Hudson. The shad have returned, striped bass regularly power upriver along the Jersey coast, and bluefish and weakfish abound. Consequently the section where the river separates New York from New Jersey is certainly one of the 100 best fishing spots in the state.

As with other large rivers, the question is not whether, but rather where to fish. Beginning in the area of the Tappan Zee Bridge, technically just above the New Jersey section, Piermont Pier offers opportunities to fish both sides of the tide in considerable comfort. By purchasing a yearly permit, anglers can drive out on the mile-long pier to fish the river. (Parking on shore is also available, but fishing then involves a bit of a walk.) Stripers and bluefish are the favorite targets here, and some very large fish are taken throughout the summer. Fish the (left) north side on outgoing tide. Move over to the right (south) side when the tide is coming in.

Tip: Throughout most of the 1990s, the best fishing occurred during the period of tide that began just as the tide was dead low and for the first two hours of incoming. Generally speaking, the river moves too quickly mid-tide, in or out. Fish the last two in and out on high water, and the first two out and in at low water.

Downriver a bit is the Alpine Boat Basin, clearly accessible from the Palisades Parkway. The boat basin provides several piers for fishing, for which pleasure a small fee is charged in the warmer months. Weakfish and fluke are caught here as well as stripers and blues.

The Englewood Boat Basin provides fine shore fishing, for massive freshwater as well as saltwater fish. Earlier in 1998, in fact, one angler caught a 25-pound carp and on his next cast, a 24-inch striped bass. The size and strength of these fish are a reminder either to hang on to your rod or put it into a sand spike that has been securely wedged into the rocks—either way with the drag wide open.

Bunker or mackerel chunks are among the best bait for the bass and blues, but many a striper is caught on bloodworms or sandworms as well. Just cut the worms into smaller pieces for the other varieties (weakfish, fluke, etc.) and put the pieces on small hooks. You may need a pyramid sinker to hold bottom, but first try a bank sinker, which is less likely to hang up in the rocks.

Further downriver, virtually right under the George Washington Bridge, is Ross Dock, which is about to undergo significant renovations as this book goes to press. When complete, it will provide a fine boat launch ramp and a long pier into the river. Meanwhile it still is a comfortable and productive place to fish.

Regarding the abundance of fish species, the river features wonderful carp fishing, from its most northern freshwater regions to the point where the water turns brackish. Big white catfish are caught up north and sturgeon are taken in lengths exceeding 30 inches. Tommy cod, a small but extremely tasty fish, are taken in great numbers, as are red hake (ling) and white perch. The revitalized shad run in the cleaner river has brought many anglers back who had deserted the Hudson years ago.

No fishing license is required on the main-stem Hudson but one is necessary on its tributaries. While the river is vastly cleaner today, it is well to remember that the New York State Department of Health has issued a health advisory that you eat no more than one meal of fish per week from any waters in New York. Filleting and skinning fish and baking or broiling the fillets (allowing the fat to run off) lessens any health risks attendant to eating fish.

■ Raritan River

The Raritan has been called the majestic queen of New Jersey rivers. This 100-mile river, the longest contained within the state, figures prominently in New Jersey's history, poetry, and natural heritage.

For anglers, the Raritan is actually three rivers, each with its own separate allures and possibilities. The South Branch, flowing from Schooley's Mountain to its confluence point, ranks among New Jersey's finest trout waters. The North Branch, classified as a Wild Trout Stream in its upper reaches, also supports fine populations of smallmouth bass, and carp. The main river, formed when the two branches meet at Branchburg, has even more to offer, including huge carp, largemouth bass, and the occasional striped bass. All three "rivers" are examined here.

24 Raritan River, South Branch

Directions: For upstream access, take Route 513 to Califon; look for side roads to the river and Fish and Game stocking signs indicating public property. For downstream access, take Route 639 north to Raritan River Road, which leads into the Ken Lockwood Gorge Wildlife Management Area.

The South Branch of the Raritan has its genesis in the hills of Morris County as an outlet from Budd Lake, flowing softly in and out of gently graded river valleys for the first few miles, then quickening its pace as it flows southeast past Schooley's Mountain through the hills to High Bridge. Ultimately the South Branch wraps around Spruce Run Reservoir and is the beneficiary of water pumped from the reservoir when the needs of the Elizabethtown Water Company require additional volume.

The South Branch gathers force and power as it flows through fabled Ken Lockwood Gorge, one of the most famous stretches of trout water to be found anywhere in the Garden State. Located north of High Bridge and east of Route 513, this 2.5-mile stretch is open to "fly fishing only" for much of the year and is closely patrolled by state conservation officers.

The South Branch may be the most heavily trout-stocked body of water in the state. For example, in a recent spring stocking season, 52,475 rainbow, brook, and brown trout were released in the upper, middle, and lower reaches of the South Branch, followed by the release of an additional 6,210 trout in the fall. As in most other good trout waters, the Raritan's trout have wide-ranging appetites. Fly fishermen match the hatch and often score well. However, the rest of us—that is, the other 99 percent of trout fishermen on the Raritan—score well with everything from pink shrimp-flavored salmon eggs and tiny meal worms to baby night crawlers, garden hackle, and floating Power Bait.

Tip: Avoid fishing on opening day of trout season at Clinton Falls, but if you have a moment, drop by with a camera. The sights and sounds of this mass of humanity flailing away at the water are unmatched anywhere we know of in the state.

Although they are both classic trout venues, the branches of the Raritan also provide a much greater variety of fishing than rainbows, brookies, and browns. Both species of bass feed actively in these waters, with smallmouths providing more than half of the action. Massive carp roll in the shallows and feed along the channel drop-offs. Indeed, the state-record carp—a beast tipping the scale at exactly 47 pounds—was caught in the South Branch in 1995 by a fortunate (and we can only assume, exhausted) sixteen-year-old angler named Billy Friedman.

Large populations of rock bass, sunfish and catfish are found in both branches. Matt Angeles caught the state-record brown bullhead in the South Branch, an impressive (for this species) 3.6 pounds. Other species, including such escapees from Spruce Run as northern pike, are regularly caught in this branch of the Raritan.

23 Raritan River, North Branch

Directions: Off Peapack Road at the old Fairgrounds, in Far Hills. Downstream, additional access is found off Route 202 (at the 202 bridge below Far Hills) and further downstream (at the Route 202–206 bridge).

The gently flowing Raritan is the largest river wholly contained in New Jersey, as it meanders more than 100 miles from the headwaters down to Raritan Bay at Perth Amboy.

Figure 3 ■ Billy Friedman with state-record 47-pound carp, caught in the
South Branch of the Raritan River.
(Photo: N.J. Division of Fish, Game, and Wildlife.)

The headwaters of the Raritan's North Branch are small brooks in Randolph Township, several of which are designated Wild Trout Streams. This area has yielded fine fishing from the time the Naritong Indians tended their fish baskets and nets set in the currents for yellow perch. The North Branch is much shorter and straighter than the South Branch, flowing through beautiful woodlands and fields to a confluence point near the town of Branchburg. (The state is mulling over the idea of building a "Confluence Reservoir" at this point, but land acquisition costs may render this unfeasible.)

On opening day of trout season, the temporary population of Branchburg jumps by 150 or more as anglers—many of them dedicated "one-day-a-year fishermen"—stand shoulder to shoulder in quest of state-stocked browns, brookies, and rainbows. The springtime popularity of the river is due in no small part to the generous stocking programs of Fish and Game, which in one recent spring stocking season placed more than 16,000 trout in the North Branch, with additional thousands stocked the following fall.

The North Branch is inaccessible for long stretches of private land holdings and consequently does not suffer much fishing pressure.

Our favorite access points are in and around the towns of North Branch and Branchburg, the latter just before the conjunction with the South Branch. ML fondly remembers an afternoon stolen from affairs of business when he took four fat trout on baby night crawlers in less than an hour, then redonned suit jacket and tie and reappeared at work with no one the wiser, unless they noted his muddy shoes.

Tip: If you can get them, try small head-hooked grass shrimp as bait in the river. Trout can't resist them.

A close friend's father had a fateful visit near this site. Near seventy and coexisting with a bad heart, this dedicated purist nonetheless fished often and expertly with fly rod and dry flies. One spring afternoon, at a point where the river flows under Route 202, he limited out and with six fine trout in his creel, started up the slight incline to his waiting car. Sadly, it was on this short walk that his heart chose to fail him. Of course, no time is a good time to die; however, for the dedicated trout fisherman who has just filled his limit, might this not be the best among bad choices?

30 Raritan River, Main Body

Directions: The Raritan is one of the most accessible bodies of water in the state, with scores of access points off Route 533 from Griggstown to Zarephath, and River Road from South Bound Brook to Highland Park. To reach Headgates, take Old York Road (Route 567) west from Route 206, past the town of Raritan, into Duke Island Park.

The Indians called the juncture of the Raritan's branches Tucca-Ramma-Hacking, or "the meeting place of the waters." At this point the converging North and South Branches form the main branch of the Raritan River, which flows "fresh" to New Brunswick, at which point tidal waters meet sweet in the shadow of Rutgers University.

After the branches merge, the river runs for a mile or two before swirling behind and over the "Headgates" in Duke Island Park in Somerset County. Since trout thrived nicely (if accidentally) in this area of clean, swift-flowing water, ML contacted Fish and Game authorities in 1973 and brought them together with the Somerset County Park Commission. Soon thereafter the main branch of the Raritan was added to the permanent stocking list for trout. (The 1997 *Fish and Game Digest* stated this area received four in-season stockings of trout, a fact which makes ML feel like a proud father!)

Finally, if trout are the most sought-after quarry in the Raritan, carp are the most challenging. Carp can be found virtually anywhere along the entire length of the river until the water turns brackish below New Brunswick. Of all these spots, however, the pool beneath the Headgates has long been our favorite, not least because the big carp here always seemed to bite in very early spring when cabin fever was its most debilitating.

One memorable day in the mid-1970s, ML was fighting an especially powerful Headgates carp when he felt the fish wedge under a submerged tree. He wanted to go in after the fish to try and work the line loose, but his waders reached only to mid-thigh and were inadequate to the task. Since RB was wearing chest waders, he volunteered to change while his partner stood holding the rod aloft to maintain pressure on the fish. In the midst of "undressing" both of them and redressing ML in his own waders, RB glanced across the wide pool at two newly arrived fishermen who were watching this spectacle and scratching their heads in wonder. (P.S. ML waded into the river, freed the fish, made it back to shore completely dry, and landed his 7-pound "buglemouth," which he promptly released.)

Tip: *If fishing Headgates, walk (carefully) through the few inches of water flowing over the waterfall ledge atop the Headgate to the side across the river for best access to carp, rock bass, and other panfish. In spring or early summer, avoid stepping on the big, ugly lamprey eels that fasten their mouths onto the covering moss of the ledge.*

Along all of its stretches—from the confluence of the two branches to the brackish water above Raritan Bay—the Raritan offers abundant fishing opportunity. Significant populations of largemouth bass, roach, sunfish, catfish, and suckers in this accessible river provide fun for anglers ranging from the neophyte to the veteran. And the fish are often large, as illustrated by the 10-pound 5-ounce state-record white catfish caught below the Route 206 bridge in Somerville in 1976 by Lewis Lomerson.

Lakes

Like diamonds scattered across a green jeweler's cloth, thousands of lakes and ponds glisten across the fertile plains and rugged mountains of New Jersey. Some 70 of the Garden State's most beautiful lakes were sculpted by retreating glaciers more than 50,000 years ago. Ten times that number are classified as natural lakes. However, the greater number of the state's 4,000 lakes and ponds have been the handiwork of man, from a handful of logs thrown across a rural stream to the stone and concrete dams of the giant reservoirs.

Anglers prize the bounty of many of these lakes, including fish that would have astounded the original settlers both in their variety and their size. The following, ranging from natural to man-made, large to small, deep to shallow, weedy to clear, are certainly among our best.

1   Greenwood Lake

Directions: Take Route 210 in upper Passaic County to access points along the lake's west bank.

Greenwood Lake came into being when debris left behind by a retreating glacier clogged an icy river called the Wanque. Early settlers examined the miles-long lake that formed behind this natural dam and dubbed it Long

Pond, the name by which it was known. Just as appropriately, they dubbed the watershed "Beautiful Valley," a name entirely appropriate to its rugged natural splendor. The icy river and the glacier carved this valley between Tuxedo Mountain on the New York shore and Bearfort Mountain on the New Jersey side, both soaring to more than 1,500 feet above the lake's pristine surface.

Lying directly on the New Jersey–New York border, the lake is claimed in part by both states. The lake was enlarged in 1768 when a colonial industrialist built a 200-foot-wide dam across the river. Another dam, constructed in 1927 in the interest of flood control, increased the size of the lake again to its current 6-mile length.

Like most mature lakes, Greenwood Lake provides an extraordinary variety of fishing opportunities. Over the years, the lake has been stocked with tens of thousands of rainbow, brown, and brook trout. In the 1990s, a great muskellunge fishery developed in the lake. (Some anglers fish Greenwood only for muskies, using monster lures and/or enormous live baits.)

However, perhaps the greatest fishing interest centers in largemouth and smallmouth bass. Both bass species grow to exceptional size, not least because of abundant food resources and superb habitat. Scores of streams and creeks winding down the mountain slopes of New York constantly renew both food stocks and oxygen in the lake, and two large islands plus the many docks and boathouses around its shore provide shelter and structure for bass.

The lake is relatively deep on average, with mean depths about 17 feet and many holes as deep as 35 feet. Because of its size, the lake provides a variety of fishing environments, ranging from heavy weedbeds in the shallows to abundant man-made and natural structure at points along the entire 6-mile length of the lake.

> Tip: *When the weather warms in late spring, concentrate on working artificial worms and buzz baits in and around the weedbeds for largemouths and smallmouths in numbers.*

Some like to recount the baseball legend that hangs over the lake. Babe Ruth, one of many celebrity anglers who favored the lake, was on his way into New York City for a World Series game when his car hit a tree on Tuxedo Mountain. The immortal "Sultan of Swat" was reduced to hitchhiking a ride to the Series. (He arrived in time to play nine full innings.)

As for us, we prefer to talk about the next trip to this wonderful old lake and about the big bass that await our offerings.

2 Shepard Lake

Directions: From Route 23 in Passaic County take Route 511 north to Sloatsburg Road. Take Morris Avenue West and then Shepard Lake Road leading north.

Nestled in the northern corner of Ringwood State Park almost on the New York border, Shepard Lake provides a variety of recreational opportunities, including what some consider the best pickerel fishing in the state. No less an authority than Ed Zendel, a very serious fisherman who operated facilities on several New Jersey lakes for many years, told us that the lake continuously produces chain pickerel in the 7- to 8-pound class, with one lunker weighing nearly 9 pounds taken in the lake in 1985.

Shepard Lake is spring-fed, with good quality cold water and enough depth to facilitate trout—especially brookies—holding over through even the hottest summers. This is good news for anglers interested in brook trout that are larger than the state's average stocked fish.

Although its average depth is more than 13 feet with some holes as deep as 30 feet, this pleasing 72-acre lake features a wide variety of fishing conditions, including rocks, boulders, lily pads, and submerged structure of various kinds. The lake's bottom substrate also varies widely, from gravel and sand to muck.

The natural bait and principal forage fish is the alewife herring, which reproduce in numbers in the cold, clear water. Curiously, the herring do not grow to significant size here; consequently, no netting of bait is practiced. This leaves more forage fish for the pickerel, trout, and other fish in the lake, which include largemouth and smallmouth bass, channel catfish, and sassy sunnies.

> *Tip: The familiar adage "big bait, big fish" does not necessarily apply here because of the smaller size of the resident herring. An occasional pickerel does succumb to a big herring, but by and large stick to smaller bait and small spinners and plugs that imitate natural conditions.*

The nearby Ringwood River, also accessible in the park, is itself a fine place to fish, especially for rainbow and brown trout that are annually stocked by Fish and Game.

A fee is charged for entry into the park on weekends from Memorial Day through Labor Day. A boat livery on the lake provides an ample number of rental boats and good launching facilities are available for trailered or car-topped boats, which can be operated with electric motors only.

Ringed by trees and heavy vegetation, Shepard Lake (a.k.a. Sheppard's Lake) is pretty, accessible, and full of fine game fish: a "consummation devoutly to be wished" by anglers interested in big fish and small waters.

8 Swartswood Lakes

Directions: Take Route 94 in Sussex County to Route 619 (East Shore Drive) northwest of Newton into Swartswood State Park.

Two fine natural lakes are located in Swartswood State Park on the Paulinskill Drainage: Big Swartswood Lake and Little Swartswood Lake. Both are extremely attractive, provide excellent fishing, and are easily accessible, with first-rate boat ramp and boat rental facilities.

The big lake, one of scores of natural basins scooped out of northwestern New Jersey by the retreating glacier 50,000 years ago, covers 494 acres with depths of up to 42 feet. Big Swartswood is a natural warm-water fishery, with good populations of largemouth bass, smallmouth bass, yellow perch, pickerel, catfish, and carp, plus swarms of several varieties of sunfish. Trout are stocked in the lake each spring and provide good fishing during April and May, and some hold over in the deepest water during summer. Occasionally walleye are taken in the lake, but bass and pickerel much more commonly take live bait offerings.

The deepest water in the lake is found along the middle of the lake's northwestern axis. Boat anglers with fish-finding electronic gear do well by double anchoring right over the drop-off at 40 feet. Shore angling is comfortable, convenient, and sometimes very productive, especially for sunfish and exceptionally large carp.

> *Tip: Crappie, yellow perch, and bass will hit live fatheads and small maribou or twister-tail jigs. Silver minnow/pork rind combinations are lethal for pickerel in summer weedlines along the lake's northern fringes.*

Little Swartswood Lake receives heavy stockings of trout. In addition, more than 2,000 tiger muskies have been stocked in this 75-acre lake over the past five years. The trout do not hold over. However, bass and tigers do well in the smaller lake, feeding on a rich diet of sunfish and minnows.

Sunfish are the most populous fish in the lake, making it an especially attractive spot to take the kids. However, representative populations of largemouths and smallmouths, pickerel and perch keep the fishing interesting for you as well.

Regarding facilities, Big Swartswood has good ramps for launching trailered or car-topped boats, plus rental boats throughout the season and live bait sold at the boathouse. Gas-powered engines are forbidden, but electric motors are allowed. Little Swartswood has no ramp but areas for launching car-topped boats are available. Seventy campsites, open year-round, are located on the big lake, with modern restrooms and hot showers making the "wilderness experience" somewhat less wild. The usual state park entrance fees are charged from Memorial Day weekend to Labor Day.

All of this, plus beautiful scenery and continuously productive fishing, make these two natural lakes the places to be, with the family or your fishing partner, through all but the coldest months of winter.

13 Wawayanda Lake

Directions: Enter the park from the east via Warwick Turnpike in Sussex County along the main entrance road.

Wawayanda State Park is a magnificent 13,000-acre mosaic of dense hardwood forests, rocky outcrops, and upland swamps. The park is located along the New York border in Sussex and Passaic Counties, and the glistening jewel set in its wild natural surroundings is Wawayanda Lake.

Clear, cold spring water from the Highland Lakes flows into Wawayanda, a gigantic impoundment capable of storing as much as two billion gallons of water for pumping into reservoirs to the south in time of drought.

So beautiful is the lake, and so peaceful, that it is almost a surprise to learn of the explosive fishing waiting along its many contours and drop-offs. The largemouth bass reigns as king of the lake, although dinner-plate-sized calico bass (crappie) and holdover trout contend for top honors. The size and growth rates of these and other fish in the lake are surprising, given the northern latitude of its location. (Most larger bass in the state are found in South Jersey, where the growing seasons are longer due to warmer average water temperatures.) The answer lies in a good population of large naturally reproducing alewife herring in Wawayanda—a population so significant that commercial netters are allowed to trap bait for resale until Memorial Day each year.

The herring provide a steady supply of forage fish for the bass and largely account for their growth. Anglers take trophy-sized largemouth bass—some as large as 9 pounds—at Wawayanda.

A combination of circumstances allows fine rainbow and brown trout to thrive in the lake: scientific stocking programs by Fish and Game, water that is both cold and clean, and depths of up to 80 feet on the lake's southern side, which helps account for the trout holding over in summer. In fact, these conditions are so productive that lake trout are being stocked here as they have been in Merrill Creek Reservoir and Round Valley Reservoir.

Tip: The deepest part of the lake is located almost dead center in the lake's southern section. Try double-anchoring along the 80-foot drop-off. Work one line just off bottom with a live herring and three-quarter-ounce sinker and fish a second live herring away from the boat on a slider float (set at a 12- to 24-foot depth, depending upon season and temperatures. This puts you in position to catch big trout, big bass, and yellow perch and, eventually, big lakers all at once).

Beside largemouth bass, crappie, and trout, the lake produces pickerel, catfish, yellow perch of formidable proportions, and scrappy sunfish, as well as the occasional smallmouth. Prehistoric-appearing bowfins, although rare in New Jersey, also are taken in Lake Wawayanda—a good 400 miles south of the Canadian waters where ML and RB have taken these strange fish.

A boathouse with a fine launching ramp for trailered or car-topped boats is located on the lake, but power is restricted to electric motors only. Canoes, rowboats, and paddle boats can be rented seven days a week in summer, and on weekends from Memorial Day to Labor Day. The usual state park entrance fees are charged and are a bargain for a day on this bountiful natural impoundment.

14 Pompton Lake

Directions: Take U.S. Route 202 in Passaic County, which parallels the lake, or the Hamburg Turnpike, which crosses its lower end at the town of Pompton Lakes.

Pompton Lake is a beautiful 204-acre impoundment fed by the Ramapo River (which changes its identity from Ramapo to Pompton below the lake). Although reasonably close to significant population centers in Passaic County, the lake is significantly underfished and offers opportunities much larger than its size or location might suggest.

These opportunities begin with largemouth bass, the number one game fish in the lake. Big largemouths are regularly taken, especially around the bridge pilings at the southern end of the lake. The largest bass hit big shiners and artificials (especially white and chartreuse spinnerbaits with gold blades).

The lake and the rivers above and below it are heavily stocked with trout. In a recent stocking season, the lake received 1,430 rainbows and browns in four spring stockings and the Ramapo more than 15,000 in the springtime alone. The Pompton River received more than 5,000 trout. Moreover, New Jersey Fish and Game stocked the lake with more than 20,000 northern pike and tiger muskie, beginning in 1989, and at least that many pike and tigers in the Pompton River. This equals big fish with plenty of teeth!

As if all this bounty weren't enough, Pompton Lake is considered one of the finest spots in New Jersey for big black crappie, led by a state-record 4-pound 8-ounce beauty weighed in by angler Andy Tintle in 1996. Moreover, the lake is a superb yellow perch spot, with many fish weighing more than a pound caught at the lake's southern end after ice-out.

> Tip: Plan to use a boat when fishing the lake, since shore-angling access is severely limited by trees and structure. Note that at the Pompton boat ramp (which needs repair) a sign shows gasoline engines up to a maximum of 9.9 h.p. are allowed but that no such limitations apply in the larger part of the lake. Also, choose summer fishing times carefully, since sunny weekends attract speedboats and Jet-Skiers.

16 Lake Hopatcong

Directions: Take exit 28 off Route I-80 to the southeastern end of the lake.

Lake Hopatcong, the largest and possibly the most popular lake in New Jersey, lies in a basin that was carved by advancing glacial ice. This beautiful natural impoundment covers 2,658 acres and runs as deep as 50 feet in spots. It provides as much variety—both of terrain and of fishing opportunity—as any still water in the state. The many bays, branches, and coves ringing the main body of the lake offer widely differing depths and fishing conditions, each more inviting than the next. One of the best spots is Brady's Bridge, especially in early spring when yellow perch are spawning. Hybrid bass are regularly taken in this area as well. Other favorite hot spots are Bertrand Island, River Styx, and Byram Cove.

Figure 4 ■ Andy Tintle with record (4.51 pounds) black crappie, caught in Pompton Lake.
(Photo: Al Ivany, N.J. Division of Fish, Game, and Wildlife.)

Places to Fish, a free publication distributed by Fish and Game, lists pickerel, catfish, yellow perch, and sunfish among the most commonly found species here, followed by hybrid bass, largemouth bass, channel catfish, smallmouth bass, and carp. Surprisingly, even this long list scarcely begins to tell the story, primarily because of the extraordinary efforts of the fabled Knee Deep Club.

Actually named the Knee Deep Hunting and Fishing Club, this club of nearly 1,000 dedicated members began stocking trout in Lake Hopatcong and other lakes in the Garden State in 1953. Knee Deepers also purchased and stocked walleye and channel catfish in the lake, and currently are adding true-strain muskellunge to the lake's rich fish population. Each year, this exceptional organization spends $20,000 or more purchasing and stocking fish, and its efforts are rewarded with ever-improving fisheries and one state-record catch after another.

Perhaps the most significant contribution of the Knee Deepers, aside from trout, was its introduction of hybrid bass to New Jersey waters. Our favorite of all fish, dubbed the "rocket" by ML because of its explosive strike and long, powerful runs, the first hybrids in the state were stocked by the club in 1985 (and subsequently in other lakes, with emphasis upon Spruce Run, by the state).

The state-record hybrid—a lunker weighing 10.14 pounds—was caught in Hopatcong in 1991 by Roy Pascoe. Equally attributable to club stocking programs was the state-record channel catfish, a 33.3-pound beast that a lucky and no doubt surprised angler named Howard Hudson pulled out of the lake in 1978.

Incidentally, although the club's name summons classic images of a fly fisherman knee deep in a pristine trout stream (or perhaps a fisherman knee deep in fish), this isn't precisely how the name was coined. According to club records, one Peg Dugan overheard the guys talking out on the porch. One of the other women asked what the men were up to and Peg Dugan replied, "I don't know, but they're knee deep in something." That "something" was the earliest planning session of one of America's classic outdoor sports organizations.

> *Tip: The main baitfish in the lake are alewife herring, which swim in massive shoals and are netted for sale to tackle stores. Largemouth bass and pickerel prey on the flashing herring and are thus especially vulnerable to shiny plugs and spinner baits. We have had particular success trolling Rapalas deep along the drop-offs, especially with big pickerel.*

Ice fishing on Hopatcong is wonderful, especially for pickerel, yellow perch, and bass. However, it can be dangerous. One or more people who have overestimated the thickness fall through nearly every year. A rule of thumb: ten consecutive days of below-freezing weather probably will produce safe ice. However, we still recommend the admittedly conservative "Luftglass Rule" (earlier co-opted from friend Dave Bank) on estimating ice safety. Simply put, it is, "If you don't see anyone out there, don't be the first to try, unless you are absolutely certain it will hold your weight."

Lake Hopatcong is a state park (with state park permit regulations applying). Walk-in fee to the park section is $1 between Memorial Day and Labor Day, and boaters pay $5 on weekdays and $7 on weekends. Fishing from shore is allowed, as is launching of car-top transported boats at various sites. Boats also can be rented from several liveries.

21 Budd Lake

Directions: Take Route 46 west of Netcong in Morris County. Lake is visible from the highway and parking available at boat ramps.

Mist rises off the water on an early summer morning and somewhere in the distance, a heavy fish rolls in the shallows. This is the pulse-quickening introduction of an ambitious early-rising angler to Budd Lake.

The accessibility of this beautiful, natural lake—right on Route 46 and close to both Routes 80 and 206—belies the excellent fishing opportunity awaiting anglers virtually twelve months a year. Because of good water quality and depths up to 14 feet, this natural 376-acre lake provides good conditions for fish growth. Consequently, Fish and Game has stocked the lake diligently over many years, with great success.

The lake is managed as a warm-water fishery, meaning that varieties other than trout are stocked on a regular basis—with special emphasis upon the pike family. Tiger muskies (crosses between northern pike and muskellunge) were first stocked in the lake in 1980, and beginning in 1981, true-strain northern pike began to be stocked as fingerlings. Both experiments have worked exceedingly well. Legal-size northerns are taken on a regular basis, and the state-record tiger muskie, a fine 18-pound 5-ounce specimen, was taken by a Budd Lake angler in 1985. (This record has since been broken.)

In all, twenty-seven species of game and forage fish have been found in the lake. At the top of the food chain are the tigers and pike, followed closely by largemouth and smallmouth bass and surprisingly large white catfish, which are a big draw to anglers interested in a tussle. The lake also holds chain pickerel, black crappie, white perch, sunfish of several varieties, carp, bullheads, and yellow perch.

ML's first visit to Budd Lake was a remarkable exercise in futility, having nothing to do with the quality of the fishing under ordinary circumstances. Accepting an invitation to fish with an experienced guide, he arrived with the barometer dropping and a storm on the horizon. Rain and hail imprisoned him in the guide's truck, and when the weather finally abated, he almost stepped on a nasty-looking copperhead snake slithering a few feet from the guide's boat. An omen? Apparently. What followed was hours on the lake with a total of four bites: one missed by the guide and three stuck and landed by ML—all catfish in this venue of big pike. A bad beginning, but fine trips on this excellent lake lay in the future.

Ice fishing on the lake is especially popular, with pickerel, yellow perch, and very large pike rewarding this chilly pursuit.

Tip: Go with shiners on tip-ups for the pike clan and occasionally work small ice jigs for perch (and to keep circulation in your hands). Don't be surprised if a big catfish takes your live bait.

There are two good boat ramps on the lake, and gasoline motors—without limitation on size—are permitted. Since the lake is shallow in spots and relatively small, boaters should minimize speed and wake.

35 Cheesequake/Hooks Creek Lake

Directions: Garden State Parkway to exit 120; follow signs to Cheesequake State Park.

Located near Raritan Bay, Cheesequake State Park straddles a transitional zone between the state's distinctly different northern and southern regions. At the southern end of the park's 1,274 acres, one encounters open fields, saltwater marshes, and white cedar swamps; at the northern end, an outstanding example of a northeastern hardwood forest. In the midst of the transition is a pretty little 10-acre lake, commonly called Cheesequake because of its location in the park. However, Lisa Barno of Fish and Game informs us that its real name is Hooks Creek Lake.

Tip: Although small, Hooks Creek Lake regularly produces excellent largemouth bass action, especially for anglers flat-lining shiners or working them a few feet below the surface.

Trout are stocked in the lake each spring and as an experiment, an additional 400 were stocked in the fall of 1997. Of course, the depth of the lake makes holding over impossible for trout, so fish are stocked with the intent of anglers taking home limits for dinner.

Channel catfish, some of considerable size, also inhabit the lake, along with large populations of sunfish. Power Bait, meal worms, garden hackle, and baby night crawlers all produce action here. No boats are allowed, but plenty of open shoreline makes fishing easy. This lake is another especially fine place to bring the kids for a few hours of fun.

The $35 State Park Pass is honored in Cheesequake State Park; otherwise, the standard $5 weekday/$7 weekend fee applies. And if you're wondering about Cheesequake, the word originated with the Lenni Lenape Indians who hunted and fished the area. It is said that Native Americans occupied this area 5,000 years before Henry Hudson sailed into Raritan Bay in 1609.

39 Farrington Lake

Directions: Take any one of four local streets (Davidson's Mill Road, Church Lane, Oakmont Avenue, or Farrington Boulevard) off U.S. Route 130 near Milltown in Middlesex County.

Farrington Lake is an enigma; a beautiful 290-acre sliver of water that remains unspoiled despite heavy population centers, including New Brunswick, East Brunswick, Milltown, and Deans, crowding around it in all directions. The lake is especially popular in the spring when scores of anglers descend on its many points of access in search of stocked trout. Fish and Game liberally rewards their quest with heavy stockings each year. Fly casters are often seen taking browns and rainbows; however, the preponderance of anglers are even more successful with natural baits, especially meal worms and small night crawlers. Power Bait also provides good results.

More seasoned anglers are apt to seek northern pike in the lake, since northerns reach considerable size here. In a netting by Fish and Game in 1997, at least one or two fish in excess of 20 pounds were taken from the lake to be milked of roe for future stocking. A number of pike in excess of 15 pounds have been taken on hook and line, and many believe that a fish larger than the record 30.2-pound northern taken in Spruce Run Reservoir plies the waters of Farrington.

Tip: To increase your odds of catching double-figure northerns, use big silver plugs (like Rebels or Rapalas) or very large shiners or herring. And be certain your hook is attached to a wire leader, since pike of any size can cut your line in an instant.

If trout get the most attention and pike are biggest, crappie are the number one fish caught in this pretty lake, especially in the spring on live bait. Fathead minnows and killies are especially lethal, the smaller the better. Rig with a small float, a tiny split shot, and a skinny wire hook like a #8 Aberdeen, fished two or three feet deep around lily pads, fallen trees, or bridges.

The lake also holds abundant stocks of catfish, largemouth bass, yellow perch, and sunfish, plus pickerel that may dart out of the lily pads

Figure 5 ▪ Northern pike being stocked by New Jersey Division of Fish,
Game, and Wildlife.
(Photo: Dave Chanda, N.J. Division of Fish, Game, and Wildlife.)

at any moment and take your live bait. Big channel catfish have also been
stocked in the lake and some very large carp are taken on corn kernels
and flavored cornmeal bait in summer.

The access points provide good spots to fish, especially around the
bridges. Several areas are packed dirt and sloped to make launching car-
toppers easy. No gasoline motors are allowed but electric motors are fine,
and all you'll ever need in this long, slender body of water.

Like most urban area fishing spots, Farrington Lake gets busy on weekends but it is nonetheless quite productive. It is an especially wonderful
place to fish with youngsters; indeed, it is wonderful for any beginner.
Abundant shoreline makes casting easy, and while we tend to focus on the
larger and more ego-enhancing fish, Farrington Lake is loaded with "kid-
sized" fish too.

40 Lefferts Lake

Directions: Take Route 34 in Monmouth County into the Borough of Matawan.

There are fishing spots which are children-friendly, great places to take the kids fishing on a warm summer afternoon. Lefferts Lake is just such a spot, with enough quality fishing to keep you fully entertained as well. This entirely accessible lake is located within the Borough of Matawan in Monmouth County. Sixty-nine acres in size and quite shallow, it usually is fished from shore, although small car-topped boats can be used if you are willing to provide the power yourself. Specifically, gas motors aren't allowed and even if they were, it is likely the boater would destroy a propeller on a hidden sunken log. (While legal, electric motors are not recommended for the same reason.)

Largemouth bass are the principal game fish in the lake, and a variety of methods will produce nice-sized fish. Our favorite artificial here is the dark (purple or black) plastic worm, rigged on a #2 weedless hook and worked very slowly just off bottom. White or chartreuse spinner baits produce bass as well. Near-black dark in summer, old standbys like the Hula Popper and the Jitterbug produce nerve-jangling hits on the surface.

Catfish and sunnies rank right along with bass both in numbers and in appetite, providing additional reasons to bring the youngsters to Lefferts Lake. Garden worms and baby night crawlers are attractive to both species, although chicken livers fished on bottom may produce more catfish.

Tip: When taking the kids fishing—especially for the first time—consider leaving your own fishing rod at home. That way, you can concentrate on being a teacher and friend, and you won't get upset if junior chucks a rock or two into the water.

While pickerel are a slow fourth in fish species, they exist in sufficient numbers and size to make fishing for them worthwhile. Killies, fathead minnows, and shiners all are pickerel killers, and you never know when a bass might take your live offering as well. Small live bait seem to work best for pickerel in this lake.

Lefferts Lake is unlikely to produce any trophy fish, but it will continue to yield hours of the kinds of pleasure that fishing is all about.

42 Amwell Lake

Directions: Take Route 31 to access road just north of County Route 579 intersection in East Amwell Township between Rocktown and Linvale. Road is marked with Fish and Game Division sign.

Armadas of towed boats and carloads of shore fishermen regularly travel past a modestly marked turnoff from Route 31 on their way to Spruce Run, Round Valley, and Merrill Creek. No doubt most of them have visions of hybrid bass, lake trout, browns, and rainbows dancing in their heads. However, when they pass the turnoff, they miss an entirely different but in some ways no less appealing fishing opportunity. They miss pretty little Amwell Lake.

The contrast to the big reservoirs is marked. Rather than hundreds of acres, Amwell Lake is slightly less than 10 acres. Rather than depths to 200 feet, Amwell Lake has a maximum depth of 12 feet. Rather than busy ramps and trailered boats, Amwell offers only (modest) car-topped boat launching and only electric motors are permitted.

On the other hand, rather than weekend crowds and sailboats and motor wakes, Amwell Lake offers peace and solitude; it is a quiet, secluded place to fish where the only "excitement" is the strike of a largemouth bass, a stocked trout, or an oversized channel catfish.

The lake was built in the early 1960s to provide flood-control protection and soil stabilization for the Stony Brook Watershed. More than a decade ago, Fish and Game began stocking with channel catfish. As a result, some lunkers cruise water around the cattails and the deeper water at the dam. More than one double-figure channel cat has been taken in the lake and many more remain. Fish and Game also has stocked the lake with largemouth bass, now the prime game fish in the lake.

And there's more. Each year, Fish and Game stocks the lake heavily with rainbow and brook trout averaging more than 10 inches in length, with a number of excess breeders up to 18 inches long added for the spring fishery.

Tip: Take boots or waders in wet weather, since some of the best shore access tends to be a bit muddy. Work small live bait around the cattails for crappie and bass.

To make the Amwell Lake experience even more convenient, abundant parking is available and the deepest water is a short cast from the dam. This neat little Hunterdon County lake has everything to recommend it.

47 Carnegie Lake

Directions: Take Route 27 south from Princeton to the lake.

Central New Jersey anglers are indebted to the generosity of the late industrialist Andrew Carnegie. In 1906, through the largesse of this famed philanthropist, the Millstone River was dammed north of Princeton to create a fine lake that is loaded with warm-water fish. This long, slender body of water, which totals 153 acres in size, actually stretches for 3.5 miles, from N.J. Route 27 east to U.S. 1.

In some respects, Carnegie is more like a river than a lake, not only in its general configuration but also in the changes that occur from end to end. At its widest point off Route 27, it is a relatively broad, flat expanse where Princeton University holds its crew regattas. Three miles away, close to Route 1, it becomes a shallow weed-choked bog. However, there is no part of Carnegie Lake that isn't pretty, no bit of shoreline that is developed, and no area that doesn't hold fish. (The lake is especially familiar to RB, who was given to stealing away from his firm's headquarters in nearby Princeton in quest of its huge carp.)

Carnegie is not stocked with trout; however, stretches of Stony Brook in Mercer County are stocked, and browns and rainbows occasionally make their way downstream into the lake. Largemouth bass are the hot fish in the lake, often growing to good size on an abundance of forage fish. Copper is an appealing color to bass in the lake, whether on the blades of spinners or buzz baits. Artificial worms rigged on weedless hooks and worked in and around the lily pads, brush piles, and other structure in the lake also produce handsomely.

The lake also is home to fine populations of crappie and pickerel, both of which are especially fond of small shiners, especially during early spring and late fall. The pickerel are more usually found in the weed-beds, while the largest crappie concentrate around submerged brush.

The biggest and certainly the most powerful fish in the lake are its carp. In the summer, it is exciting to watch them roll in the shallows. They are quite wary and require considerable skill to catch in numbers. An angler from England (where carp are ranked the number two game fish behind salmon) demonstrated a unique and, to us, unappealing style of fishing at Carnegie Lake, which involved the use of a slingshot and packages of live maggots. He launched one slingshot load after another into the lake for chum, then baited his hooks with the offensive little creatures and cast them out among the chum. To our surprise, the carp were less finicky than we were. In fact, he caught an impressive mess of fish. Since slingshots are illegal in New Jersey and maggots are culturally taboo, his system is not likely to catch on here.

Tip: *When fishing for carp here or in any other waters, fish carefully selected bait flat on bottom and keep your reel drag loose or your reel bail open. Otherwise a big carp will own your favorite spinning outfit.*

Big channel cats also roam the shallow waters of the lake and can be taken on a wide variety of baits, including shiners intended for bass and pickerel.

There is a boat ramp for launching small boats but only electric motors are permitted. There is also plenty of shoreline to fish on this super accessible lake, with especially good fishing near the dam.

52 Lake Mercer

Directions: Take Route 1 east of I-295 to Quakerbridge Road (Route 533) into Mercer County Park.

Located in a highly developed county park within a stone's throw of Mercer County Community College, Lake Mercer does not provide quite the same wilderness feeling as some of New Jersey's more remotely situated lakes. However, it offers largemouth bass and pickerel fishing that rivals the best.

Much of the shoreline of this pretty 275-acre lake is ringed with brush, both in and out of the water. This provides good habitat for bass and other game fish, including pickerel, crappie, and yellow perch. Extremely large channel catfish and carp also lurk in the shallows.

The lake reaches depths of 20 feet and has been scientifically stocked by Fish and Game with a variety of species over a period of years, including more than 12,000 tiger muskies between 1989 and 1996. However, reports of these fish coming to the net are few.

An occasional smallmouth bass is taken in the lake but largemouths reign supreme. Large shiners on floats work very well virtually all year long. Long stretches of brush and other structure around the shoreline invite the boating angler to throw weedless artificial worms, Power Worms, and Slug-gos almost on shore, then slowly retrieve the bait through the structure. This produces fine action in the lake, especially in spring and early fall. There are some steep drop-offs that also hold bass on their edges in the deeper parts of the lake and shiner "look-alikes" like Rapala and Rebel lures get their attention.

There are good populations of pickerel, channel catfish, crappie, and sunfish in the lake, and smaller populations of carp and yellow perch.

Tip: The channel catfish are particularly large and offer a fine scrap if caught on light gear. Even if your primary interest is bass fishing, try taking along a few chicken livers or hunks of cut mackerel. Fish your bait dead on bottom—with your drag open.

Mercer County maintains a marina with a first-rate boat ramp on the lake for trailered and car-topped boats, plus rowboat rentals. Gasoline motors are not permitted. In addition, nearby facilities include everything from tennis to biking and handicapped trails, plus a nearby ice-skating rink (not on the lake). All in all, this is one of the nicest places in the state to bring the family for a few hours of pure pleasure.

54 Assunpink Lake

Directions: I-195 to exit 11 (Imlaystown Road) and head north on a paved road that turns into dirt before reaching Assunpink Lake.

The name Assunpink usually suggests one lake; in fact, it should suggest three, all located in the beautiful and bountiful Assunpink Wildlife Management Area (WMA) in Monmouth County. This wildlife management area, and especially the smaller Stone Tavern and Rising Sun Lakes, represents a triumph for the joint efforts of the U.S. Soil Conservation Service, the New Jersey Division of Fish, Game, and Wildlife, and the New Jersey Green Acres Program. These agencies worked together to acquire land along the Assunpink Creek drainage to provide flood protection, fish and wildlife habitat, and recreation. Their efforts were wildly successful.

The largest and shallowest of the three lakes in this 5,000-acre wildlife area is the namesake Lake Assunpink, covering more than 225 acres with only 12 feet of water at its deepest point. This lake was stocked by the state with hybrid striped bass, an experiment that has not been notably successful—although a few thoroughly shocked anglers hook these "rockets" each year. The problem is that most of the hybrids discovered an outlet (Assunpink Creek) and swam out of the lake, eventually winding up in the Delaware River. However, largemouth bass are a different story. Big bass are commonly taken at the lake, especially in early spring. (Most are photographed and released.) Spinner baits, plastic worms, and buzz baits produce best when cast from boats into structure and weed growth along the shoreline.

Tip: Try "flat-lining" live bait here, working large shiners on light line with no float or sinker. Flat-lined baits flutter at the surface and attract bass of all stripes.

Assunpink Lake also is a first-rate spot for crappie and pickerel. Big numbers of crappie are caught by slow trolling of small weedless spoons and by working small shiners, fathead minnows, or killies around structure. (If you are fishing for the table, remember to check your compendium.

At this writing, the limit is ten fish at least 8 inches in length, rather than the more usual twenty-five panfish limit.) There may be as many chain pickerel in Assunpink as bass. Live bait works best for these sharp-toothed game fish. Unlike northern pike, pickerel are best fished for without wire leaders. The wire will definitely cut down on the number of strikes, by a factor of two to three times. (You will lose the occasional fish to cut leader, but the vast majority of pickerel will be hooked in the side of the jaw.)

Hard-fighting channel catfish, stocked by the state many years ago and now grown to weights in double numbers, also provide great fishing fun for those who aren't finicky about bait. Ripe chicken intestines, cut (and aged) mackerel or herring, chunks of liver or store-bought "stink" baits all attract these sleek, powerful channel cats (as well as the indigenous and pesky stocks of yellow catfish).

Sunfish and a few yellow perch fill out the bill in Assunpink Lake, the former more commonly taken on garden hackle or baby night crawlers; the latter on very small shiners.

55 Rising Sun Lake

Directions: Take Route 571 into the Assunpink Creek WMA in Monmouth County. Access to Rising Sun Lake is right on 571.

The second and smallest of the Assunpink Creek WMA lakes is Rising Sun Lake. Accessible right from the road with a convenient parking lot for shore anglers, this picturesque lake covers 38 acres and has depths of up to 20 feet.

This lake is often overlooked by anglers lured by larger waters; however, it has good resident populations of largemouth bass, pickerel, crappie, and sunfish, and oversized channel catfish in numbers. In addition, according to Ed Booth, who operates a fine boat-rental facility at Assunpink Lake, it has schools of yellow perch, which offer two possibilities; the large perch are fine table fare, the smallest are excellent bait.

Tip: *If you happen to catch a very small yellow perch, rig him on a second line fitted out with a float and a size #2 blue Aberdeen hook and chuck him out to work for you. After a short wait, a big pickerel will often invite himself to dinner.*

In Rising Sun Lake, a boat ramp is available for both car-topped and trailered boats. Only electric motors are permitted.

56 Stone Tavern Lake

Directions: Take I-195 in Monmouth County to Route 571 into the Assunpink Creek WMA.

Stone Tavern Lake, a third lake in the Assunpink Creek WMA, is a fine alternate for anglers looking for solitude on weekends when crowds descend on the larger Assunpink Lake. With depths up to 24 feet, this 52-acre lake provides good fishing in summer when all of the shallower, warmer lakes have stopped producing.

Stone Tavern was stocked with largemouth bass and bluegills in 1971; consequently, resident populations of both fish are large and feisty. Artificials worked deep are usually the ticket here for largemouths, although live bait produces its share of fine keeper fish. There are common bullheads and yellow perch in Stone Tavern, but not in sufficient numbers to annoy the bait fisherman questing for larger fish.

The largest fish in the lake are channel catfish, many exceeding 10 pounds.

Tip: *The rule on "whiskerfish" bait is the same here as in Assunpink Lake; that is, the less attractive the bait is to you, the more attractive it is likely to be to the channels. Also, big bait equals big channel cats, so try a whole chicken liver (on a #2 bait holder hook, fished dead on bottom). You might be in for some great fun.*

70 Mirror Lake

Directions: Take Route 70 West in Burlington County to Route 530, directly into the town of Browns Mills.

Mirror Lake is a fine example of Fish and Game providing for the fishing appetites of our children and ourselves in years to come. This pretty

250-acre lake is right in the heart of Browns Mills Township, just off Main Street in the middle of town. Part of the lake is privately owned, part municipally owned. Much of it is accessible and angler-friendly, with good parking at a new boat ramp built in 1995 for public use.

For many years the lake was chock full of sunfish, bullheads, and carp. However, in 1992, the dam on Main Street needed significant repairs, necessitating significant draining of the lake. Before this operation began, a four-day state salvage operation was undertaken to save the resident fish population. Half of the rescued fish were removed to nearby Pemberton Lake. The rest were stocked into Little Pine and Big Pine Lakes.

According to super source Hugh Carberry of Fish and Game, after repairs were completed, the lake was restocked with a highly desirable mix of largemouth bass, black crappie, channel catfish, pickerel, and yellow perch. In 1997, Fish and Game electroshocked the lake and found that the stocked populations had prospered nicely, with keeper-sized largemouths and a fine population of pickerel at the top of the food chain. (The fast growth of the fish is illustrated particularly well by a 25-inch pickerel caught in mid-January 1998, by an angler casting a very small jig.)

Tip: Drift small minnows on floats around weeds and structure for crappie but fish deeper for yellow perch. Cut slivers of dead bait for chum, especially for perch.

Mirror Lake drains into the North Branch of Rancocas Creek, itself rich in fishing opportunities from the lake's outfall to the main stem of the Rancocas and ultimately the Delaware River. Shore fishing is permitted where lake property is not privately owned, and as noted, first-rate boat-launching facilities are provided. All in all, Fish and Game has made a real commitment to making Mirror Lake a first-rate fishery, and as in so many other lakes and streams throughout the state, we are all the direct beneficiaries.

75 D.O.D. Lake

Directions: Take Route 130 in Salem County to Pedricktown and the lake.

Named the D.O.D. (for Department of Defense), this pretty lake was owned by the Army Corps of Engineers. Its original purpose was as unglamorous as its name: it was to be a dump site for dredge materials. Fortunately, it was never used for this purpose, since this would have spoiled a truly outstanding fishing lake. (Significantly, D.O.D. is being taken over by Fish and Game, which undoubtedly will improve the fishing even more.)

The hourglass-shaped lake is some 130 acres in size, making it one of the largest bodies of fresh water in all of South Jersey. It is spring-fed and there is foot access to a tidal gate to the Delaware River at its southwest corner.

Hugh Carberry of Fish and Game shared his encyclopedic knowledge of freshwater fishing in South Jersey with us, in discussing D.O.D. and several other lakes and rivers. He told us that largemouth bass weighing 9 pounds have been caught in D.O.D. and that it is quite possible that a state-record "bucketmouth" swims in these little-known waters. He added that his department had turned up three primary species of fish in several electroshock and gill-net samplings of the lake. First in interest to anglers are bass, many of them quite large. Second, and certainly of great interest to the authors, are great numbers of extremely large carp, some weighing as much as 30 pounds. Third—smaller in size but huge in numbers and fishing fun—are white perch.

Entrance to the lake had been restricted by the Army Corps of Engineers to a gate, which opened only from 8 a.m. to 5 p.m. daily. Presumably this limited access will change. Of course, in cold weather, these are the prime hours anyway.

Tip: When fishing cold water, go at noon when the water has started to warm up a little. If it is a breezy but sunny day, fish with the wind blowing into your face, since the warmer surface water will be blown downwind, pushing the best bass fishing toward you.

The fishing is good both from boat and shore. A small dirt ramp located at the neck of the hourglass provides sufficient room to launch a good-sized boat. (Chances are the State will install a fine concrete ramp once the lake is under its control.) Car-toppers can launch all around the lake with relative ease. Shore anglers can pick their spots since there are not too many trees at water's edge.

81 Harrisonville Lake

Directions: Drive south on Route 45 from Mullica Hill to Route 617. Take 617 to Avis Mill Road to the lake.

Harrisonville Lake is a relatively small body of water with surprisingly diverse and often exciting fishing possibilities. Located in the Harrisonville Lake WMA right on the line between Gloucester and Salem Counties, the lake is generously trout-stocked by Fish and Game (more than 1,600 trout were stocked in a recent spring stocking season). And while none hold over through the hot months of summer, they provide opportunities for great fishing pleasure and tasty dinners for many South Jersey anglers and their families. The state also recently stocked big channel catfish in the lake, some larger than 10 pounds, according to our source at Larry's Fisherman's Cove in Mantua (who notes the fishing is best in the lake during the cooler months of spring and fall).

According to the most current issue of *Places to Fish,* the state gives its highest rating (very good) to the largemouth bass, sunfish, and (our personal favorite) carp fishing here. It also notes good populations of pickerel, yellow perch, and catfish.

> *Tip: When fishing for trout, try "popping up" meal worms by rigging with a miniature yellow marshmallow pushed up to the eye of the hook and then the mealie. Rig with a split shot or two a foot above the hook and cast out with a small float. The mallow will pop up your bait off bottom where all cruising trout can see it. (If a carp happens to take your bait, be prepared for an incredible fight.)*

Largemouth bass here prefer the largest roach (shiners) you can find. The old adage "big bait, big fish" works especially well here, since the bass are accustomed to feeding on stocked trout. When the water is cooler and at dark, surface poppers work well here, as do weedless white and/or chartreuse spinner baits. The lake's abundant pickerel, yellow perch, and crappie prefer smaller shiners.

Most anglers fish from shore, although car-toppers (electric motors only) are legal at this inviting and productive little lake.

82 Iona Lake

Directions: Take Route 40 East in Gloucester County to Route 613 to the lakefront.

Iona Lake is another of South Jersey's many surprises for freshwater anglers. Located in the center of Gloucester County, this 36-acre lake offers fine trout fishing in spring and fall plus super bass fishing virtually year-rou.nd, with trophy pickerel thrown in for good measure.

The lake features a creek channel and a great deal of fish-friendly habitat, including oxygen-rich water from the feeder creek, brush piles, little islands, and long stretches of coontail grass and other weeds. It is heavily stocked with trout by Fish and Game each spring and fall, providing hours of angling pleasure for spin casters and fly anglers alike.

Tip: One of the best, and certainly the most convenient, baits for trout, in Iona Lake and in almost any other trout water, is canned corn. We have probably taken more smallish trout on Green Giant kernels rigged on small hooks and fished on or near bottom than any other single bait we use.

The trout in this clear-water lake are particularly obliging to shore-bound anglers since they tend to congregate in deeper water, which is within easy casting distance of the accessible lakefront. These hatchery-raised fish are susceptible to the full range of baits that are productive in other waters, from corn (as noted) to baby night crawlers, salmon eggs, and bits of cheese.

Largemouth bass and pickerel also abound in the lake and are caught virtually year-round. In summer, successful anglers work the deeper water along the creek bed, often with lures of the most unlikely hues (including chartreuse, bright yellow, and bright pink). Because of limitations on area and forage fish, the bass and pickerel do not grow rapidly in the lake, generally averaging between 1 and 1.5 pounds. However, enough larger fish—some exceedingly large—are taken each year to keep the fishing exciting. Indeed, the state-record chain pickerel was taken on the flats at the back of the lake. The lake also is loaded with bluegills.

A boat launch is available on the east side of the stretch along Route 613. While some private properties restrict shore fishing, there is plenty of convenient shoreline access.

86 Parvin Lake

Directions: South onto Parvin Mill Road from Route 540 in Salem County, about 6 miles west of Vineland, into Parvin State Park.

Parvin Lake offers the angler a beautiful trip into a natural area of oak, pine, and swamp hardwood forests, glimpses of rare species of birds and animals, and excellent fishing, all for the price of a $1 walk-in fee in season. (The $35 State Park Pass eliminates even this small price for a truly wonderful experience.)

Located in the heart of a 1,125-acre state park on one of the tributaries of the Maurice River, this 99-acre lake is the largest in Salem County. It is shallow, with a mean depth of less than 4 feet. Consequently the lake tends to be unproductive in mid-summer, when the water is too warm and often muddy. However, most of the rest of the year produces surprisingly good fishing for a wide variety of species—with special emphasis upon largemouth bass. In fact, Parvin is designated as a "trophy bass" lake, which means anglers may keep no more than three bass daily, and none less than 15 inches in length.

Bass are by no means the only fish in this pretty, man-made lake. Common bullhead catfish and sunnies flash through the shallows in exceptional numbers. Pickerel, and white and yellow perch also await the angler.

Tip: If fishing artificials, stay shallow, using shallow-diving crankbaits and large surface plugs worked around the weedbeds and stumps. Anything that runs deep is almost sure to be lost in submerged tree stumps and snags.

Big carp roll in the shallows of this lake but take special skill to land because of bottom structure. If fishing for these "big scalers," use prepared carp bait or corn kernels and be sure to leave your bail open or your drag wide open. Otherwise your rod and reel are in severe danger of being lost to a powerful strike and run.

Parvin Lake catfish can be taken on a variety of baits and provide a good scrap on light tackle. Exercise care in removing hooks—in your best interest and the fish's. If the hook is swallowed, simply cut the line and release the fish. Any wrenching at the line will probably kill the fish, whereas the undisturbed swallowed hook, in all likelihood, will not. Exercise care in unhooking even the lip-hooked catfish, since all of this species have sharp fin tips that are slightly poisonous and easily jabbed into careless fingers.

Shore fishing is allowed, but we prefer working the lake from an electric (only) motor-powered car-topper. Launch facilities leave a good deal to be desired, but there is a boat livery on the lake. Catch a trophy bass at Parvin but remember the rules and regulations, including minimum size limits and closed seasons for these fine shallow-water fish.

87 Lenape Lake

Directions: Take Atlantic City Expressway to Route 50 south in Atlantic County, to town of May's Landing and lake.

Lenape Lake provides anglers with a two-for-one opportunity that is much to be prized. The lake itself is a wonderful place to fish. In addition, it is located at the Great Egg Harbor River, which offers its own outstanding brackish-water angling opportunities. (Especially fine fishing can be found at a very large bulkhead in the river at what was called Spooney's Marina.) Thus if the lake happens to be slow, a few steps take the angler to the river and an entirely different (and perhaps more productive) experience.

Our own experience with Lenape Lake has been supplemented handsomely by Hugh Carberry, who handles Warmwater Fisheries Research for Fish and Game. Hugh checks out the lake regularly and reports particular success with pickerel, using smaller Mepp's spinners (in sizes #1 and #2) and floating plastic worms. (Golden shiners, the main forage fish in the lake, are great natural bait in the lake.)

Lenape also features fine populations of slab-sized black crappie, yellow perch, bucketmouth bass, bullhead catfish, and sunnies.

Tip: *There is a large "run" of spawning yellow perch each spring to the deeper waters around the dam. Use grass shrimp or small shiners, either with a float that holds the bait just off bottom or with bottom rigs.*

The water in Lenape is characteristically dark. It has little vegetation but plenty of wood to provide structure and fish habitat. At 350 acres, it is the largest lake in Atlantic County, with plenty of shoreline and boat-ramp angling. Launching facilities are available and gasoline motors up to 9.9 h.p. are allowed. Ice fishing is generally not an option since the surface of the lake seldom freezes sufficiently to provide "safe ice." However, in an unusually bitter winter, this might be an interesting option, given the lake's large populations of winter-feeding perch and pickerel.

88 Union Lake

Directions: Take Route 55 in Cumberland County just northwest of Millville; exit into Union Lake WMA.

Union Lake is southern New Jersey's largest freshwater body of water and certainly one of its hottest fishing spots. Originally built more than 200 years ago, the lake and its surrounding property were sold to private interests and the dam reconstructed in 1868. In 1982, through the auspices of the Green Acres Program, the lake was purchased by New Jersey Division of Fish, Game, and Wildlife, which set about making it a brilliant success as a managed fishery.

The 898-acre lake is situated along the Maurice River, a location that has much to do with its skyrocketing reputation as a fishing hot spot. The reason is the fish ladder installed by the state in 1991, which permitted large numbers of river herring and gizzard shad to enter the lake from the river. The alewife herring and gizzard shad have remained in the lake and are reproducing prolifically. This increased base of forage fish has been the foundation for an explosion of bass growth and bass fishing, which just keeps getting better.

Union Lake holds four species of bass: largemouths, smallmouths, true striped bass, and hybrid bass, although the latter species has been noticeably absent from "creel counts." Of course, hybrid bass are noted migrants and it is likely that many stocked fish have escaped into the Maurice River via the dam at Millville. The populations of smallmouths and stripers both are relatively limited, but the lake is one of the most active largemouth bass fisheries in the state. Also present in numbers in the lake are pickerel, sunfish, crappie, common catfish, white perch, and carp. Moreover, the state began experimental stocking of tiger muskie in 1996.

Grass shrimp are absolutely deadly on perch, crappie, and sunnies. Artificial worms and Slug-gos produce big results in the reed grass, and deep-running crank baits work wonders around the lake's great variety of fishable structure.

Tip: Try catching very small sunfish on a tiny hook and bit of worm to use as bait for pickerel and largemouth bass. Fish three-inch sunnies around structure, either "flat-lined" or with a medium float and splitshot.

Union Lake also is the site of a significant freshwater artificial reef program, in which old tires and evergreen trees are weighted and placed in the lake to improve fish habitat. To our knowledge, artificial reefs have been "planted" in at least eight sites—all marked with 15-inch red buoys. The quality of fishing around these artificial structures attests to their effectiveness.

It should be noted that a 1993 assessment of mercury concentrations in Union Lake species resulted in an advisory against eating largemouth bass, chain pickerel, and yellow bullheads from the lake. This is, in our view, not altogether a bad thing, since fish that have yielded anglers all of the pleasure of the strike and fight are released to fight again another day.

Two boat launches are available: a fine state-of-the-art facility constructed by Fish and Game in 1993–94 (at the west side of the lake) and another operated by the city of Millville at the southeastern end. Outboards of up to 9.9 h.p. are allowed in this gem of a South Jersey lake.

Reservoirs

The fountainheads of New Jersey—each an individual engineering marvel constructed to slake the thirst of New Jersey during summer—are the reservoirs. Ranging in size from 25 to 1,290 acres and in depth from 20 to more than 200 feet, these beautiful impoundments have been virtual laboratories for Fish and Game's stocking efforts, with spectacular results. Six of the most impressive state-record fish have come from three of these reservoirs: Round Valley, Spruce Run, and Monksville. Our own experience includes catches of as many as 71 trout (caught and carefully released) in a single day. There are simply no superlatives that exaggerate the fishing opportunities they provide. Following are some of the best.

4 Monksville Reservoir

Directions: Take Route 511 past Ringwood Boro, north of Wanaque, in Passaic County.

Monksville Reservoir is a beautiful, deep-water impoundment situated in the new Long Pond Iron Works State Park, just south of the New York border in Passaic County. The reservoir covers a total of 505 acres and reaches depths of 90 feet. It is regularly stocked with trout as is its feeder river, and most anglers come for the trout. The lake also is a fine fishery for smallmouth bass, largemouth bass, walleye, and pickerel. The walleye fishery is especially promising, thanks to the Federal Aid to Sportfish Restoration Program working with New Jersey Fish and Game. A survey conducted in spring of 1996 indicated a walleye population approaching 3,000, with fish averaging more than 19 inches in length.

However, it is the muskellunge fishing that has generated the most excitement at Monksville, especially since an angler named Bob Neals shattered the state record in late January of 1997 when he pulled a 42-pound 8-ounce muskie through the ice. Fishing at the north side of the lake in an area of about 18 feet of water, Bob had taken the precaution of using a braided wire leader to protect against sharp teeth, just in case. The muskie hit a live shiner, tripped one of the tip-ups, and waged a ferocious battle, ultimately straightening two small hand gaffs the angler tried to use to pull the massive fish through the ice. Finally a heavier gaff got the job done and Neals prepared to release the fish back into the lake when another angler urged him to have it weighed on a certified scale in the likely event it was a state record. Good advice! Measuring 48 inches in length, with a girth of 26.5 inches, this fish may keep Neals in the books for quite a while.

The story of this and other muskellunge fisheries in New Jersey traces to 1983 when a small group of avid muskie anglers formed a chapter of a national organization called Muskies Inc. The group raised funds and with the approval of Fish and Game, stocked Greenwood Lake with 300 fingerlings in 1985. The group continued to stock this fine lake each year, and within five years legal-sized muskellunge were being caught. During the same period, a muskie fishery began to develop in Monksville Reservoir, although no fish were stocked there. There is evidence that these fish escaped Greenwood Lake at a time of heavy rains and made the short trip downriver via the connecting Wanaque River.

In 1995, a survey indicated that that keepers taken in Monksville Reservoir averaged 39.5 inches in length.

Figure 6 ■ Bob Neals shattered the state record with this 42.13-pound muskellunge caught in Monksville Reservoir.
(Photo: Debra Elmire, N.J. Division of Fish, Game, and Wildlife.)

Tip: Large bait does indeed equal large fish in the case of muskies. Fish live bait of 9 inches or more in length if you can find them for this voracious fish, any time of year.

There are two good boat ramps on the lake with plenty of parking. No entrance fee is charged for using the ramps, parking, or any other facilities in Long Pond Iron Works State Park.

6 Canistear Reservoir

Directions: Route 23 in Passaic County to Canistear Road, which leads directly to parking and boat ramp.

Canistear is one of several fine reservoirs owned and carefully managed by the Newark Watershed Conservation and Development Corporation (NWCDC), all located in the Pequannock Watershed in Passaic County. Over the years, fishing and boating have been very carefully controlled, including many seasons closed to all but local residents holding permits. As a result, this beautiful reservoir and others managed by the NWCDC are underfished and incredibly rich with opportunity for nonresidents who may now obtain permits.

It is thought that state-record largemouth bass, smallmouth bass, and pickerel may be caught in Canistear and its sister impoundment, Clinton Reservoir. And while the reservoirs are now open to broader use, it is clear that the NWCDC will continue to control boating and fishing in the reservoir. This control, together with bountiful stocking programs, assures wonderful angling opportunities for years to come.

Canistear is a roughly oval-shaped lake of 350 acres, with maximum depths exceeding 40 feet. The lake features all of the warm-water sport fish plus occasional trout. The structure and flora of the lake favor bass and pickerel, with shallow, weedy coves leading out to steep drop-offs and deeper water. In summer, the largest bass tend to feed just at dawn and from dusk to full black of night, and are especially susceptible to large artificial worms. However, many anglers have been fooled into believing that massive bass are feeding in the dark of night by what were in fact huge carp rolling in the shallows.

> *Tip:* *Locate the deepest water with a depth finder and double-anchor as close as possible to the 42-foot depth (directly out from the dam). Set a slider float at a depth that corresponds to the thermocline and cast it along the drop-off line, baited with a large shiner. Work a second line at bottom with a live crawfish for smallmouths or small shiner for yellow perch.*

Pickerel in large numbers are found in the weedy shallows and are caught on plugs that simulate herring or large shiners.

The fishing and boating season at Canistear opens in early April and runs through mid-October. However, this reservoir and the others in the watershed may be closed during drought conditions.

As noted, special permits are required for use of the NWCDC reservoirs and at this writing, anglers must make launching reservations. Moreover, all persons who fish the Newark Watershed must have a valid State of New Jersey fishing license. Gasoline motors are prohibited.

A boat ramp is provided just off Canistear Road at about the reservoir's mid-point. The best fishing from shore is available on the west bank, which requires a bit of a hike from the parking area. It is, however, a walk that is well worth the effort.

7 Clinton Reservoir

Directions: Take Route 23 in Passaic County; turn right onto Clinton Road just north of Newfoundland.

Clinton Reservoir, a second Newark Watershed impoundment, is noted for extremely large bass and pickerel. In 1975, the first year in which public fishing was allowed in the watershed reservoirs, an 8-pound 8-ounce largemouth was caught in this fine lake, as was a 6-pound 2-ounce chain pickerel. Not long afterward, a 5-pound 8-ounce smallmouth also was taken here. It boggles the mind to consider fish this large, more than twenty growing seasons ago.

The reservoir is 423 acres in size and runs to a maximum depth of 47 feet. It wraps two arms around Buck Mountain and one of these arms connects via a shallow, swampy area to Buckabear Pond. The weed-choked shallows provide an ideal nursery for small pickerel, which hide in the weeds and devour a seemingly unending supply of minnows until they graduate to feeding on small bass and perch in the lake. This protected beginning and the abundant food resources in the deeper water help explain the presence of so many large pickerel in the reservoir.

Tip: Pickerel feed along the weedline edges and are especially susceptible to crank baits and spinnerbaits cast into the weeds. The larger bass, however, tend to be found around more permanent structures, including massive boulders easily visible from the boat dock.

Large populations of yellow perch and black crappie also feed on the large supply of minnows and would attract more angling attention were Clinton not such a superb bass fishery. Trout also are heavily stocked in the reservoir—3,640 rainbows, browns, and brook trout in a recent stocking season. Significantly, trout hold over in Clinton. Consequently, lunker trout swim these waters alongside exceptionally large bass and pickerel.

Parking and boat-launching facilities are available on the reservoir. The number of boats is rigidly controlled. Anglers are advised to call to make launching reservations (973) 697–2850, especially during drought conditions. As at Canistear, special restrictions make it advisable to call before making a trip.

15 Haledon Reservoir

Directions: Take route 208 in Passaic County to Ewing Avenue in North Haledon, to the reservoir.

One of the best-kept secrets about super angling spots in New Jersey is the Haledon Reservoir. At 25 acres in size with depths ranging to 20 feet, this picturesque little impoundment is fed by springs and a brook that keep the water cool, clean, and well oxygenated. In the past, it could be fished (by special permit) only by anglers of three nearby towns (North Haledon, Haledon, and Prospect Park). Consequently it appears on no fishing maps and is seldom discussed even by the most loquacious outdoor writers. As of 1998, however, the public can apply for and receive permits, most likely for a fee. We believe Haledon Reservoir will thus be a hot topic and a fishing spot you should know.

Perhaps the most knowledgeable authority on the reservoir's bountiful opportunities is Dave of nearby Bates Guns and Tackle. Dave, who fishes the reservoir regularly, characterizes largemouth bass as the number one fish in the lake, based on size and explosiveness of strike and fight. However, for sheer numbers and fishing fun, yellow perch contend for top honors. Catches of 100 yellow perch in a day here are not unusual.

Tip: Fish deep for perch, using small shiners. And don't let the abundance of fish cause you to forget that 25 per day is the current legal maximum to take home.

No boats are allowed on the reservoir. However, excellent access, heightened recently with a major shoreline cleanup, makes angling from shore entirely comfortable. There is quite a lot of rocky bottom (suggesting largemouth bass) and blowdowns and heavy weeds, providing habitat for the reservoir's abundant bass, black crappie, and pickerel populations. Big pickerel (Dave has landed a 24-incher and had two as large as 30 inches cut him off at his feet) show particular interest in black and gold Rapalas plus the more usual small to medium shiners.

The reservoir also holds giant carp and plenty of catfish, too.

Recognize the two old guys waiting in line ahead of you for permits to this Reservoir? Yes, it's us!

25 Merrill Creek Reservoir

Directions: Exit 17 from Route 78, follow route 31 north for 10.5 miles, follow Route 57 (also East Washington Street) 6.3 miles to Montana Road. Follow signs to lake and boat ramp.

Merrill Creek Reservoir is so beautiful, so perfect, that the angler's heart tends to skip a beat when the water first comes into view. Unlike other large lakes in New Jersey, this 650-acre impoundment is situated on top of a mountain. Consequently it is exposed to almost no runoff and the water quality is excellent.

The fish population is exceptional, due largely to the efforts of Fish and Game. First and foremost, the lake is home to thousands of lake trout, courtesy of annual fall state stockings and strictly monitored creel and

Figure 7 ■ New Jersey Division of Fish, Game, and Wildlife sample netting of lake trout at Merrill Creek Reservoir.
(Photo: N.J. Division of Fish, Game, and Wildlife.)

size limits, plus significant populations of brown and rainbow trout.
Largemouth and smallmouth bass are also present in good numbers,
many lurking amid the standing timber that still rings the young reservoir's shorelines. There also are walleyes, yellow perch, and other panfish. Another denizen of the lake is the lamprey eel, which accounts for
the occasional experience of landing a trout with a nasty scar on its body.

Because of its relative age and, until recently, a shortage of natural
forage fish, Merrill Creek has produced no record fish as of this writing,
although several bass to 7 pounds and rainbow trout to 6 pounds have
been caught. But the number of fish—especially trout—that can be caught
(and released) in a single day is close to mind-boggling. For example, the
first five times ML fished the lake—in the fall of 1995—he had only a con-
tour map to guide him and a basic idea of how to fish deep water (prac-

ticed often at Round Valley Reservoir). Despite his lack of knowledge of the lake, he landed and released a dozen smallmouth bass, two rainbows, thirty brownies, and well over one hundred lake trout in those five trips. And this was with all of the inefficiencies of learning the water as he went along. Fishing together in 1996 and 1997, we had many single days of catching and releasing between fifty and seventy lakers and browns.

Now the state is working to increase the forage-fish population. Because the predators had eaten virtually all of the natural alewife herring bait, the growth rate of trout and bass had slowed. Consequently, Fish and Game purchased and released tens of thousands of herring into the reservoir in 1997. In addition, a temporary change in the law as of 1998 permitted two lakers of 15 inches or more to be kept—the same size limit as brown trout and rainbows. (Heretofore, the legal laker limit was one fish of 24 inches or more.) In no circumstances can lakers be kept if caught between September 16 through November 30, so as to allow them to spawn.

> *Tip:* *Despite the numbers of hungry fish, just showing up with gear and bait won't guarantee a catch. Successful fishing requires effort and skill. ML's method for catching trout in deep water personifies that combination. Here's how it works. Find a significant slope with your depth finder in roughly 100 feet of water. Double-anchor your boat at 100 plus feet at one end and 80 feet at the other, with at least 275 feet of anchor line in the deeper water and 200 feet at the shallower end. With the boat anchored tight over 85 to 90 feet of water, suspend a shiner or herring at 15 to 25 feet for brownies and rainbows. Then cut up some bait in French-cut slices and throw them over for chum. (Yes, chum for trout!) Tie a #6 model #3906 Mustad Sproat hook onto a 3-foot leader below your black-barrel swivel, and drop it to the bottom with a three-quarter-ounce egg sinker above the barrel. Experiment with baiting styles. A herring, either head-hooked or hooked behind the dorsal works and held just off bottom, works well. A large shiner hooked behind the dorsal and fished on bottom with a few inches of slack line (Uncle Nick method) usually produces even better laker action.*

Since the odds favor undersized lakers, make sure to reel them up slowly; otherwise they can get the "bends" and die. Also keep a sharp pair of scissors handy to cut off hooks buried in the fish's throat or belly. Fish can live nicely with a hook gradually dissolving in their throats; however, a hook wrenched out of a fish's entrails guarantees that you will release a corpse.

Regulations governing Merrill Creek are different, and in some cases unique, from other lakes in the state. For example, your fishing boat must be at least 12 feet long. Boaters may not be on the water before daylight or after dark. Gasoline engines are absolutely illegal. They may be on your boat's transom (beside your legal electric motor) but must be out of the water. No swimming is allowed. No tubers, scuba divers, or other distractions to fishing are allowed either.

All of these regulations tend to keep Merrill Creek a perfect fishing experience: a spectacularly beautiful lake, plenty of hungry fish, lots of peace and quiet. Incidentally, when driving toward the parking lot, you will see a road sign which reads, "No Outlet." Given what fishing means to both of us, we find this sign particularly amusing. After all, what better outlet exists on earth than this wonderful mountain lake, so far removed from traffic and executive responsibilities and the pressures of everyday life.

26 Spruce Run Reservoir

Directions: Exit 17 off Route 78 to Route 31 north in Hunterdon County. Follow 31 north approximately 4 miles; turn left at Van Syckle's Corner Road traffic light. Go 1.5 miles to park entrance on your left.

Of all the waters in New Jersey, salt or sweet, still or running, this gorgeous 1,290-acre impoundment is our favorite fishing spot. Our affection for Spruce Run began early and has continued unabated. In fact, ML was in line with his little 8-foot Viking rowboat at 1:30 a.m. on the first day in 1965 that boats were permitted access to the lake—behind forty other lunatics! He caught three nice Donaldson trout that day. A week later, RB landed his first keeper rainbow trout in Spruce Run, another fat Donaldson.

Located at the outfall of Spuce Run Creek on Route 31, the reservoir's construction was finished in 1964 with the completion of the main dam and two small dikes. Almost immediately the state began scientific game-fish stocking programs, initially featuring only largemouth bass and rainbow trout of the Donaldson strain.

Conditions seemed to favor trout, especially since the reservoir is continuously fed by four streams except in drought conditions. Two of these streams, Mulhockaway Creek and Spruce Run, provide spawning and nursery areas for brown trout. However, despite depths of up to 70 feet and good supplies of fresh water in the spring and winter, the lake loses most of its oxygen supply below a depth of 10 to 15 feet each summer when water levels are usually lowest. Because the top layer holding the oxygen becomes too warm for trout, most cannot hold over except for the enterprising brownies that hold in and around the cooler, oxygen-rich streams.

However, most other species have no such problems and consequently largemouth bass, pike, smallmouth bass, calico bass (a.k.a. crappie), and hybrid bass abound by the thousands, foraging on abundant stocks of alewife herring and golden shiners. Yellow perch and three varieties of catfish (common bullheads or yellow cats, white catfish, and the more recently stocked channel catfish) also are caught in numbers, with a few of the channel cats having grown larger than the fine 8-pounder ML caught and released in 1997.

A whopping 117,942 northern pike have been stocked in Spruce Run from 1978 to the date of this writing, and the fishery for northerns is correspondingly excellent. German carp find the reservoir much to their liking, growing to mammoth size and sending the hearts (and boats) of novice bass fishermen racing as these fish roll in the shallows.

Spruce Run is one of a handful of fishing spots in New Jersey that offer 365 days a year of angling fun. In cold winters, the lake generally freezes solid and ice fishermen catch yellow perch, largemouth bass, and monster northern pike.

In the spring, when the last ice is out, recently stocked trout attract most interest, especially on Opening Day, when throngs of anglers flail

Figure 8 ■ Manny with a 5-pound Spruce Run "rocket" (hybrid bass).
(*Photo: Manny Luftglass.*)

the water to a froth, especially at the mouth of Spruce Run Creek. At the
same time, however, log-sized northern pike are feeding in the shallows.

In later spring and early summer, big crappie, largemouth bass, yellow perch, and then spectacular hybrid bass begin to feed in earnest. Big
carp and catfish also get serious as the water warms.

On calm days in the summer, lucky anglers may be treated to an especially thrilling experience. It begins with diving gulls, reminiscent of
birds at the shore powering down into bluefish feeding frenzies. The surface of the water suddenly comes alive as entire shoals of frantic alewife
herring flutter and dart ahead of slashing hybrid bass. If you are anchored,
as we often are, and the fluttering comes toward your boat, you need to

get herring into its path, flat-lined or on relatively shallow slider floats, or both. Four wildly fighting hybrids hooked at the same time is not an unheard-of consequence. (If you are underway, of course, the trick is to chase the action and cast just ahead of the feeding fish.) Either way, you may be in for a few minutes of action that will be the high point of an entire summer's fishing.

In the fall, northern pike and largemouth bass which have been finicky feeders in the hottest months rejoin the action, along with the hybrids and a few holdover trout. Fall is our favorite time to fish Spruce Run because the average hybrid bass and pike are bigger and the yellow perch, a favorite fish for the table, are fatter. Slab-sized crappie start feeding as well. Most importantly, the sailboats and most of the anglers are off the water, leaving things pretty much to us two cranky old citizens, competing hard as always for the biggest and the most fish caught (and almost invariably, most fish released).

Most fish of all varieties in Spruce Run are taken on live bait, although a number of successful bass anglers stick to plugs, trolling or casting to likely spots. The fishing is best from boats, but a few places offer super angling for shore fishermen, especially for stocked trout, pike, carp, and yellow perch.

As to hours, the fish in Spruce Run tend to be forgiving of anglers who don't rise before dawn. Of course, very early or very late is still best during the warm months of summer. But we have taken fish at all hours of the day and night, especially when the water isn't overly warm. A glance at the "Solunar Tables" is always recommended for the best fishing time.

Tip: *For hybrid bass, no time in summer is better than "magic hour," the hour period between daylight and 30 minutes before black dark. Sitting double-anchored over a drop-off with live herring working on slider floats and waiting for a heart-stopping, ripping scream of drag is an experience you absolutely owe yourself.*

As to depths and places, shallow water early each year as temperatures start to climb and deeper water in the summer months, but with live bait fished only ten to 15 feet down to stay alive in the oxygen layer. Fish deeper in early spring and late fall, especially for yellow perch and catfish.

As to spots, the creek mouths are good for trout in springtime and the drop-offs near Campers Point are good most times the lake isn't too crowded. The Power Lines is a good area early or late each year for hybrids, crappie, and largemouth bass, and Goose Island offers good bassing opportunities too. The rocky shore at the dam is good for bass and as noted, Black Brook Cove is a first-rate area for big northerns.

A boat-rental facility is open for most of the fishing season, featuring rowboats with or without motors, as well as sailboats and paddleboats— all for a modest fee. While outside the scope of this book, camping, picnicking, swimming (at the lifeguard-protected sandy swimming beach), and sailing are popular activities in the park.

A final note: driving through the park and down to the boat-launch site, especially early in the morning, can be a wonderful wildlife experience all by itself. Does and beautiful spotted fawns regularly feed along the roadways in spring and summer. An occasional fox can be seen skulking through the early morning shadows. Flights of mallards, black ducks, and bright puddle ducks circle overhead. The woods ring with the sounds of pheasant, bobwhite quail, and doves. The eerie cries of loons are heard on the water. And if you look carefully, you might be fortunate enough to see the resident eagle flying high over the water in search of a careless trout.

28 Round Valley Reservoir

Directions: Take Route 22 in Hunterdon County; turn at Round Valley Recreation area signs just south of Lebanon and follow road to lake.

Round Valley is commonly referred to as "The Valley of the Giants" and with good reason. No body of water in New Jersey other than the Delaware River has produced so many record fish. At this writing, the state-record brown trout, lake trout, smallmouth bass, and American eel all were taken in Round Valley. The sizes of each of these record fish provides only a hint of what awaits the serious angler in Round Valley. The brown

Figure 9 ■ Carl Bird with his state-record (24.85 pounds) Round Valley
lake trout.
(Photo: Al Ivany, N.J. Division of Fish, Game, and Wildlife.)

trout, caught by Lenny Saccente on a live herring, weighed 21.6 pounds.
The laker, which hit Carl Bird's lure, was even larger at 24.1 pounds. The
"bronzeback," caught in 1990 by Carol Marciniak, weighed in at a hefty
7.2 pounds, exactly one pound more than the record eel caught by James
Long in 1994.

Built as a backup to nearby Spruce Run Reservoir for water supply
in severe droughts or in case dam repairs draw down Spruce Run too dras-
tically, the beautiful and unspoiled Round Valley is the second largest body
of fresh water in the Garden State (second only to Lake Hopatcong).
Opened in 1972, this Hunterdon County impoundment, when full, covers
2,350 acres of land with 55 billion gallons of blue water going as deep as
175 feet.

Beyond question, Round Valley is the finest fishing venue in the
state for very large trout. For example, ML caught 17 lakers exceeding 24
inches in the first six months of 1997 alone. In the same year, the authors

caught no less than 60 brown and rainbow trout exceeding 15 inches in length, with the largest a fine 23.5-inch brown. It should be noted that Round Valley is designated a "trophy trout lake," which means that size and bag limits are significantly different from most other waters in the state. In 1998, for example, an angler was allowed to keep two brown and/or rainbow trout if the fish were at least 15 inches in length and one lake trout of at least 24 inches. (Lakers were out of season during spawning time from September 16 to October 30.)

Round Valley is another wonderful success story for the New Jersey Department of Fish, Game, and Wildlife, which has created a superb fishery with its scientifically managed stocking programs of brown trout, rainbows, and lakers. We should add that the Round Valley Trout Association, a great fishing club numbering more than 1,000 members, is dedicated to the quality of fishing in the lake, enthusiastically raising money to buy and stock trout in the lake as well. Their efforts are so closely allied to those of Fish and Game that members assist Fish and Game when it stocks the lake.

Significantly, Round Valley is the only lake in New Jersey containing breeding trout. Its rainbow and brown trout cannot reproduce in still water, but lake trout have spawned and reproduced in significant numbers, virtually eliminating the need for Fish and Game to stock more.

Although much of the fishing focus at Round Valley is on trout, the lake supports fine populations of largemouth and smallmouth bass, yellow perch, and sunfish. The panfish can be caught in numbers close to shore, providing excellent table fare and a perfect early fishing experience for young children. A population of pickerel also has mysteriously made its way into the lake and, like other species, its members have grown to exceptional proportions. An 8-pound chain pickerel was taken several years ago and ML caught one topping the 5-pound mark early one December morning.

Successful fishing methods vary widely at Round Valley. Because of its great depth and trophy-sized game fish, many anglers prefer to troll using downriggers to keep plugs or spoons running deep. Others prefer to drift, often "jigging" heavy metal jigs just off bottom or "trolling" live bait. Shore fishermen catch large numbers of stocked rainbows and browns in cold water, often no more than 100 yards from the boat ramp. Our preferred method is double-anchoring our boat and fishing live shiners or herring, suspended at depths where we "read" fish or at bottom for lakers. While anchored, we generally fish other lines higher up—flat-lined or with slider floats—for rainbows and browns.

> *Tip:* *Freeze your unused shiners or herring from earlier fishing trips and bring them along on ice. Some of the frozen bait will float while others will drop to the bottom of your bait pail. When fishing for lakers, try the RB method of hooking a floating herring about halfway between tail and dorsal fin and dropping it down with a three-quarter-ounce barrel sinker just dusting bottom. The herring will "swim" above the sinker in a natural looking upright attitude, thus attracting lake trout.*

Facilities include a fine boat ramp available to the public at no charge and plenty of parking. The maximum outboard motor size is 9.9 horsepower and you must have a Coast Guard–approved life jacket (cushions are not acceptable) for each person in the boat. Like fish size and bag limits, these rules are strictly enforced, keeping Round Valley clean, pristine, and safe.

60 Manasquan Reservoir

Directions: Take Route I-195 West to exit 28B for Route 9 north, Freehold, in Monmouth County. Turn right at first traffic light on Georgia Tavern Road. Go 0.3 miles; turn right into Windeler Road to Reservoir.

Manasquan Reservoir is one of the youngest impoundments in the state and certainly one of the most promising fisheries in Monmouth County.

Before this beautiful 770-acre reservoir was filled, a number of fish-concentration devices and gravel spawning areas were installed by Fish and Game. In addition, 108 acres of standing timber and 28 acres of stumps (complete with log-brush shelters) were left for fish habitat.

Although Fish and Game has concentrated primarily on stocking warm-water species of game and forage fish, some rainbow and brook trout are stocked each spring (and are thought to hold over in the 40-foot depths). In addition to trout, nine species have been stocked since 1990, among them smallmouth bass, largemouth bass, bluegill sunfish, black crappie, and hybrid bass, plus alewife herring and fathead minnows for forage.

Of great additional interest, the reservoir is in the developmental stages of a hybrid bass stocking program that is already producing stellar results, due to superb conditions for growth. When stocked in 1994, the hybrids averaged 4 inches in length. A year later, they had grown to 10.7 inches on average, and by 1997, many of these "rockets" had reached the minimum legal size of 16 inches, and they continue to grow on a rich diet of alewife herring.

In fact, all of the stocking and management programs are producing excellent results in this new reservoir. Anglers are regularly taking 3- to 4-pound bass (largemouths and smallmouths) plus plenty of keeper crappie and bluegills.

> *Tip: Cast crankbaits and diving plugs around the standing timber and submerged structure for bass. Work live bait around the timber, flat-lined or with slider floats, for largemouths and crappie.*

Facilities at Manasquan Reservoir include a fishing pier, floating dock, two boat-launch areas, and excellent parking, plus a bait and tackle shop at the Joseph C. Irvin Recreational Area Visitor Center, which is located at the southeast corner of the reservoir.

Smaller Freshwater Rivers

New Jersey is defined by flowing water. From fresh inland rivers and streams to salty tidal tributaries racing to the sea, some moving waterway flows through or past every one of New Jersey's 567 municipalities. Our eastern and western borders are the mighty Hudson and Delaware Rivers. Thus most of us live with the soft songs and pleasures of streams or rivers close at hand.

For anglers, the rivers and streams of New Jersey are a constant source of fascination, challenge, and excitement. In the smaller rivers discussed here, a particular deep channel, a long stretch of rapids, an undercut bank, a fallen tree, a deep, quiet pool live in our memories and call out to us to return.

3 Upper Wanaque River

Directions: Take Route 511 in Morris County west of Ringwood to East Shore Road, which runs through the Wildlife Management Area (WMA) to the river.

The Upper Wanaque River is a picturesque tree-shaded stream flowing from Greenwood Lake downstream to Monksville Reservoir. This section of the river, just over 2 miles in length, flows largely through the 2,277-acre Wanaque Wildlife Management Area in upper Passaic County. This WMA is a virtual outdoor paradise, with significant populations of wild turkey, whitetail deer, and many varieties of upland game available to hunters and a massive 45-acre pond loaded with largemouth bass and chain pickerel. However, many feel that the jewel of this outdoor paradise is the river.

The special attraction in the Upper Wanaque is trout. Each spring and fall, Fish and Game stocks the river heavily with rainbows, browns, and brookies, and when water conditions are favorable, extremely large fish are taken, some as heavy as 6 pounds.

This stretch of the Wanaque is a virtual game-fish highway between the two massive reservoirs at its head and foot. It is the route taken by true-strain muskellunge originally stocked in Greenwood Lake, but escaping via the river to establish a sensational muskie fishery in Monksville Reservoir. It is the route taken by walleye upriver from the lower reservoir during the spring spawn. No doubt some of the larger trout similarly managed to escape the confines of still water.

> Tip: *If you want to fish the Upper Wanaque, leave your fly rod at home. Dense overhanging trees along the accessible stretches of the river make fly casting nearly impossible. Bring live bait, preferably small shiners and baby night crawlers, and fish the deepest holes.*

Access is directly from East Shore Road, along which small parking areas have been provided for the convenience and safety of anglers. No fee is charged for using any of the WMA's facilities.

Fishing the Upper Wanaque in summer may be a problem because of low water conditions. Early spring and late fall are the best times to come in search of the river's abundant populations of trout and other game fish. And if water conditions aren't just right and the fishing is a bit slow, Green Turtle Pond is at your elbow and loaded with largemouth bass and pickerel.

5 Big Flatbrook

Directions: Exit Route 206 in Sussex County, turn west onto Route 521, then onto Route 515 through Flatbrook-Roy WMA. For mid-section access, use County Route 521. For lower section take Route 615 to Wallpack WMA.

The Big Flatbrook is one of the classic trout streams in the eastern United States, rivaling the Beaverkill in New York and Penn's Creek in Central Pennsylvania for rugged beauty and superb fishing. Running some 30 miles from its headwaters high in the Kittatinny Mountains, it provides every possible condition conducive to trout, from rock-strewn rapids to deep, shaded pools; from fast riffles and pocket water to classic rock-and-gravel-bottom runs. Anglers longing for the spiritual experience of fly fishing seen in the movie *A River Runs Through It* can find the rugged beauty, the peacefulness, and the shattering excitement all at once when fishing the Big Flatbrook.

Figure 10 ■ Colorado mountain stream? Nope, it's the beautiful Flatbrook in northern New Jersey.
(Photo: Al Ivany, N.J. Division of Fish, Game, and Wildlife.)

For more pedestrian reasons, it is difficult to capture this wonderful stream in one piece, because it changes so significantly from stretch to stretch. For example, from the river's origins for roughly a third of its length, the Big Flatbrook is a picturesque mountain stream, providing wonderful trout fishing in April and May—especially for the spinning-tackle devotee equipped with worms and talent. It would be natural to assume that this first and most pristine stretch is best; however, this is not the case. It is in the middle stretch of the stream turned river, commencing at the Route 206 bridge and continuing downward to the junction with the Little Flatbrook—a fine trout stream in its own right—that the finest fishing is to be found. The reason is that at this point the stream has been cooled and aerated by numerous springs and tributaries as it tumbles through the mountains and is oxygen-rich and food-rich to promote trout growth. Numerous species of minnows and a wide variety of insects provide an unending source of food for the fish and a variety of natural baits for the fly caster to emulate. Ultra-clean water and gravel bottoms contribute to the success of the holdover trout population. Another significant contribution to the quality of the fishing is the state's stocking program, which favored the Big Flatbrook with more than 30,000 adult trout in one recent season.

A 4-mile stretch of the river is designated "fly fishing only" and a portion is a "no-kill" area. Native brook trout and stream-bred browns are found here in numbers. However, they are wary and provide a test of the fly caster's skill and experience.

The lower portion of the river, from Walpack to the Delaware River, is no less beautiful than its upper reaches nor any less promising. In fact, the last several miles of the river are in many ways the most attractive. Averaging 60 feet in width with holes as deep as 6 feet, the river is still gin clear and highly productive of trout. However, it also boasts a first-rate population of smallmouth bass and a variety of panfish, plus whatever large fish may travel up this clear, bait-rich stream from the big river. Bait, lures, and flies all are permitted, with small crawfish and hellgramites especially deadly for smallmouths.

Tip: Monster striped bass are often seen on top early on summer mornings, just above the river's junction with the Delaware. They are elusive and difficult to catch, especially since no one who knows how to catch them is willing to talk.

A day on any portion of the Big Flatbrook almost guarantees that you will be back. Such fine fishing in so much natural beauty assures your return.

9 Paulinskill River

Directions: New Jersey Route 94 parallels the upper reaches of the river from Newton to Columbia in Sussex County. Look for side roads to the water's edge. The mid-section of the river is accessible along Route 15 in Lafayette or Paulinskill Road (off Route 94), which parallels the river. The lower section is accessible via county Routes 610 and 521.

The Paulinskill is a river of changing moods and opportunities. It flows virtually its entire course in the shadow of the rugged Kittatinny Mountain Range. In many spots these rugged hills actually extend to the water's edge. The Lenni Lenapes fished, hunted, and trapped along virtually the full length of the Paulinskill and governed their people from villages along its shores.

The headwaters of the Paulinskill trickle out of the hills north of Newton, calling to mind spring-fed mountain streams in the neighboring Pennsylvania mountains. A mile or so below, the water is backed up by an old dam, and from there the little river cascades through rocky ravines that send the trout angler's heart racing and fingers digging through fly boxes for the perfect offering. The wildest and most beautiful scenery and the fastest water may be found in the few miles from Augusta to Paulins Kill.

Tip: Privately owned property blocks access to much of this spectacular water. However, keep your eyes open as you drive along the river. At signs marking the sections stocked by the state, access to both the river and the state's largesse will be available.

The second stretch, from Marksboro downriver for about 2.5 miles to Paulina, is by far the most heavily fished section of the river, which broadens to 60 feet or more and features fast water, slow deep pools, many productive riffles, and 500 yards of spectacular rapids. This stretch of the river provides prime trout and smallmouth bass fishing when the water is cool, primarily in early fall and from early spring through the warming trends in late June. Conditions make it ideal for spin casting and fly casting, bait and artificials. Favorite baits include baby night crawlers, garden hackle, minnows, and salmon eggs. The most productive lures here are very small spinners and a variety of nymphs. In warmer weather, sunfish, rock bass, chub, and catfish keep the angler busy. (The largest "brown trout" ML ever caught here turned out to be a log-length sucker!)

Parking and good access can be found along stocked stretches of the river. For those wading, care must be exercised because of soft, silty bottoms.

From Paulina to the river's juncture with the Delaware, the river is accessible and easy to fish. This section of the river warms earlier than its upper reaches and therefore trout anglers tend to fish it earlier in spring and to desert it earlier in summer. They also concentrate on broad, deep pools where larger fish hold in the current.

Although the river broadens and slows in its lower reaches, it still has surprises in store. After flattening through a big bend above Hainesburg, it suddenly constricts between high, sheer cliffs that produce more than a mile of fast rapids which summon images of the wildest wilderness streams. From the foot of these rapids, it flows through beautiful meadow country to its meeting with the Delaware.

The Paulinskill is heavily stocked by Fish and Game, with more than 29,000 trout released in the spring stocking in a recent year. Thus extraordinary throngs of fishermen flock to its banks on opening day. However, for most of the rest of the season, the entire length of this graceful stream is underfished by all but knowledgeable anglers who understand and appreciate its bounty.

10 Ramapo River

Directions: The Ramapo is paralleled by Route 202 for virtually its entire flow through New Jersey. Access points from Route 202 include the Route 17 bridge near Mahwah and Ramapo Valley Road north of Oakland, in the Ramapo Valley County Reservation.

Anglers hungering for wild, rugged beauty and fine trout fishing are never disappointed with trips to the Ramapo River. The countryside surrounding long stretches of the river features majestic forests and looming mountains, in some places crowding right down to the water's edge.

Originating in New York State, the river's New Jersey run is only about 9 miles before it flows into Pompton Lake. The river continues below the lake, with a changed character and name—becoming the Pompton River as it flows south toward its juncture with the Passaic.

The final several miles of the river above the lake are the most popular with anglers, not least because of its easy accessibility. However, it is by no means the most satisfying fishing experience. In our view, the best fishing for trout and smallmouth bass is still found in deep holes along the river's upper reaches, including the area around the Route 17 bridge.

The river is heavily stocked with rainbow, brook, and brown trout each year and many anglers concentrate almost solely on trout. However, they often are surprised—not unpleasantly—to take smallmouth and largemouth bass, especially on live bait offerings and small silver spinners. In addition, the river supports healthy populations of chain pickerel and black crappie. The non-trout species are especially susceptible to shiners and crawfish, and while the bass tend to be on the small side, they provide fine sport on light tackle.

The character and nature of the river, however, suggests trout. At its upper reaches, the Ramapo has fine, deep pools alternating with gentle rapids and undercut banks where trout hold in the current. Due to good dissolved oxygen in this cold, clean stream in summer, many trout hold over, so that the possibility of a large fish looms with every cast.

Tip: *Shrimp-flavored pink salmon eggs are often the most productive bait in the river.*

Although housing and commercial development are now encroaching on the Ramapo Valley, a trip to the river's more remote sections still provides a fine experience for the angler bent on a solitary evening along a wild steam.

17 Pequest River

Directions: U.S. Route 46 parallels miles of the lower stretch of the river in Sussex County. Excellent access and parking in the Pequest WMA, from Route 46 east of the town of Pequest.

The Pequest is one of New Jersey's natural treasures; a beautiful little river with exceptionally large native trout in its riffles and pools. Ranging from 30 to 50 feet in width, the Pequest flows through long boulder-strewn stretches and features deep pools, riffles that often dimple and flash with rising trout and pocket waters that make the trout angler hurry preparations to get on the water.

Significantly, this river's most productive fishing spots also are its most accessible: its mid- and lower stretches—often paralleled by U.S. Route 46—beginning around Great Meadows and ending with its confluence with the Delaware. Through miles of rocky pools and riffles, the angler fishes pristine waters reminiscent of trout waters in the high mountains of the American west.

The lower river supports significant populations of smallmouth bass and a variety of panfish. However, the Pequest is treasured by anglers primarily as a trout stream and with excellent reason. It produces browns and rainbows of 6 pounds or more, and late summer evenings provide the breathtaking view of numerous trout rising to hatching aquatic insects. Moreover, its fine populations of wild trout are regularly supplemented with significant stockings of rainbows, browns, and brookies.

Figure 11 ■ Marianne Hochswinder with opening-day (1991) Pequest
River trout.
(Photo: Al Ivany, N.J. Division of Fish, Game, and Wildlife.)

Tip: While fly casting may appeal to the aesthetic senses of anglers, light spin casting with bait produces most of the fish. Worms are most effective, bounced along the rocky bottom or riffles to emulate natural foods in the current. Tiny shiners also are productive in some of the deeper pools. The emphasis in this wild trout environment is "natural" presentation.

An added attraction is the state-of-the-art fish hatchery on the Pequest, which is worth a visit by anglers interested in the outstanding work of Fish and Game. Since beginning operation in 1983, the hatchery has consistently met its goals of producing 600,000 quality trout. For example, one spring stocking placed 604,000 trout from the Pequest Hatchery in almost 200 lakes and streams throughout the state, producing fine fisheries for rainbow, brown, and brook trout. To enhance fishing opportunities even more, a fall program was added in which the state stocked over 100,000 additional production-sized (10 inches average) trout plus 1,200 excess rainbow trout broodstock averaging 18 inches.

A visit to the hatchery can accomplish two purposes: first, the angler gets a fine view of the outstanding work of Fish and Game; second, the Pequest WMA not only provides good access and easy parking but it marks the beginning of the best fishing on the river.

It should be noted that the river is designated a Seasonal Trout Conservation Area for about a mile downriver from the hatchery. After late May, anglers can keep only one trout at least 15 inches in length in this regulated area.

From the hatchery downriver to the Delaware, the Pequest provides fine fishing opportunities. However, early in the season, anglers will have to work to distance themselves from growing throngs who also have discovered the exceptional fishing.

Izaak Walton, the early chronicler of fishing, wrote, "I love any discourse of rivers, and fish and fishing." The beautiful Pequest would surely have gladdened his soul.

18 Rockaway River

Directions: Take Union Turnpike (Route 15) in Morris County south of Mt. Arlington to Berkshire Valley WMA for upper river access. Midsections of the river are accessed via Route 513 near Dover and Route 46 south of Denville.

The Rockaway River is one of New Jersey's great conservation success stories; a river on the brink of destruction that was reborn when concerned citizens decided to fight for its life. With fifteen urban communities and scores of manufacturing facilities located along its banks, the viability of the Rockaway was severely threatened by pollution and carelessness. But a group called the Friends of the Rockaway River came together to lead the good fight to reclaim their river. Eventually they enlisted the aid of a dozen groups ranging from Trout Unlimited and the American Rivers Association to the U.S. Fish and Wildlife Service, and their success has been close to legendary.

Today, the Rockaway is a beautiful, healthy river that is home to fine populations of native trout and other fish species. The state-record brook trout, a 7-pound 3-ounce battler, was caught in the Rockaway by Andrew DuJack in 1995. This is one of many large trout to be found along most of the river's entire length.

The beauty of the Rockaway is enhanced by continuous changes in its personality. In its upper regions, the river is as narrow as 20 feet, flowing first through long, quiet riffles, then suddenly gathering force before slowing again. Its broader middle section includes flat, deep pools, boulder-strewn stretches, and 2 miles of spectacular water below the falls and gorge just below Boonton. In addition, no less than a dozen dams change the character and personality of this little river.

The Rockaway is considered one of the most underrated trout fishing streams in the state, and with good reason. Flowing no more than 20 miles from New York City and through a long succession of towns and communities, the river seems an illogical place to catch trout. However, the Rockaway is the exception that proves the rule that appearances can be deceiving.

Access to the best fishing in the river's upper reaches is found in the Berkshire Valley WMA, south of the town of Mt. Arlington. Regular stocking programs keep the section of the river flowing through this tract an attractive place to spend a spring morning, and no fees are charged for use of the WMA.

The most productive section of the river is also one of the most easily accessible (directly from a loop of State Highway #513 between Wharton and Dover, and also along part of the Morris Canal towpath). Just under 2 miles of this section have been classified as Trout Maintenance Waters. Not surprisingly, holdover trout are caught in this section, and it was here that the state-record brookie was taken.

Several fine spots are found right in the town of Boonton, where a nearby sporting-goods store stocks trophy-sized trout each year and awards prizes for specially tagged fish.

Tip: *The best color bait here, as in several other New Jersey streams, is pink. Use pink Power Bait or pink shrimp-flavored salmon eggs for fine results.*

The final section of the river supports outstanding populations of smallmouth and largemouth bass, catfish, carp, and sunfish, all there for the taking of urban anglers with a yen for a lovely outdoor experience.

19 Passaic River

Directions: For access to the best trout fishing, take I-287 to North Maple Avenue in Basking Ridge (in Morris County) and turn east into Lord Stirling Park. Or take Route 24 in Chatham to Fairmont Avenue to River Road for Passaic River Park access.

The Great Swamp, a surviving remnant of ancient Lake Passaic and the Wisconsin Glacier, drains into the sinuous Passaic River, which borders or traverses seven counties on its route to Newark Bay. Along much of the river's upper length, areas of tangled swampland and marshes isolate the angler from the bustle of densely populated northern New Jersey, recalling earlier, simpler times. The headwaters of the river produce a rich diversity of wildlife, from the familiar whitetail deer, muskrat, and fox to more than 200 species of birds and classes of flora and fauna seen almost nowhere else in the region.

The Great Falls of the Passaic River is a natural wonder created 200 million years ago, when underground disturbances pushed molten rock up through the earth's surface. These heat-formed rocks make up the chasm and cliffs of the falls area, where a billion gallons of water flow over the 77-foot-high falls each day. So dramatic are the falls that Dutch missionaries are said to have described it as "a sight to be seen in order to observe the power and wonder of God."

In 1798, *Dobson's Encyclopedia* noted that the Passaic "abounds with fish of various kinds." That description remains true to this date. The stretch of the river from the Great Swamp down to Mount Bethel is heavily stocked with trout by Fish and Game. Moreover, long stretches of the river are owned by Somerset County and thus provide "hassle-free" access to good fishing along the shore. The river also features excellent populations of largemouth bass, catfish, and carp.

Many miles downriver, the Dundee Dam area at Garfield features some of the most diverse and exciting fishing on the river, including superb striped bass fishing with lots of "linesides" making their way upstream from Newark Bay.

Tip: Go with bucktails for stripers in this area.

The Dundee Dam impoundment also produces extraordinary carp fishing. A fellow member of the Carp Anglers Group of America fished the Passaic at this point with a friend in 1997. Using floating bread for bait (a system much more popular in the United Kingdom than the United States) they landed and released 23 carp up to 14 pounds in six hours.

To add to the diversity and challenge along this stretch of the river, New Jersey Fish and Game stocked tiger muskies in 1987.

The Passaic is New Jersey's second longest river, flowing 85 miles from the Great Swamp to Newark Bay through the largest and most heavily industrialized cities in the state. It is a testimonial to the combined efforts of environmentalists and federal, state, and county governments that it remains a productive fishery along every mile of its looping path to the sea.

20 Musconetcong River

Directions: Route 31 in Sussex County, to Hampton and local roads to the river.

For thousands of anglers the name Musconetcong summons thoughts of trout fishing. Some travel from out of state to ply its pristine 40-mile flow, and with good reason. The state stocks the "Musky" more heavily than almost any other river or stream, often with more than 50,000 trout in a single stocking season, and the fishing is superb.

The Musconetcong River actually begins as a modest rivulet flowing into Lake Hopatcong. The outflow of Hopatcong runs a bit more than a mile to (and through) the river's namesake impoundment, Lake Musconetcong. Thereafter the "Musky" flows almost straight for 40 miles to its appointment with the Delaware River at Riegelsville.

The river's original name was Musconetkonk, thought to mean "place where bass fish are caught with spears" in original Munsi Indian dialect. The river is as pleasing to fishermen as it is frustrating to canoeists because of the hundreds of fishways and dams installed to provide good fishing waters and for flood control. It is thought by some that more work has been done for fish propagation along the Musconetcong than along any other river in New Jersey.

Fly casters are enthusiastic about the great "hatches" on the Musconetcong throughout the season, from big tan caddis hatches in May through late-season hatches of whiteflies in early September. There is no section of this fine river that is not hospitable to trout, from the nutrient-rich effluent of the lake to the lower end near Asbury, where many springs supply the river with cool, clean water.

A handful of private clubs control stretches of the river, especially along its lower reaches. However, the state has done a fine job acquiring access rights to long stretches of the river. Among the most accessible are found at Hampton Borough Park, Stephens State Park (just upriver from Hackettstown), and along New Jersey Route 57 from Penwell to Hackettstown. The Woodstream Rod and Gun Club of Highland Park held annual competitions on the river every year at Butler Park near Hampton. A member of the club named Manny Ettlinger displayed incredible skill with silver or gold size #3 Mepps spinners each year, catching trout after trout. (Another club member named Manny has a 19-inch smallmouth bass on his wall, which took a silver #3 one cool spring morning.)

Tip: Use a silver spinner on bright days and a gold spinner when skies are overcast. When fishing for bass and trout in clean river waters, avoid feathers or fuzz.

22 Black River

Directions: Take Route 24 southwest of Chester in Morris County to Hacklebarney State Park Road. Follow road into park.

The romantic notion of rivers springing out of rugged mountain wildernesses and flowing through virginal woodlands into increasingly developed areas is reversed by the Black River. Indeed, the Black's headwater streams spring up behind shopping malls and heavily traveled roads around Succasunna. As the river flows south, it reaches increasingly wild and beautiful country, especially at its midpoint around Hacklebarney State Park.

The river flows through Black River Park in Chester Township and the massive Black River WMA northeast of Chester. On paper this would seem to be the logical place to begin fishing, since the river is accessible along publicly owned land. However, the fishing is relatively poor in these shallow, swampy upper reaches, which are littered with felled trees, logjams, and decaying vegetation.

Rinehard Brook and Trout Brook, both wild trout streams and the Black's principal tributaries, empty into the river at Hacklebarney State Park. This is one of the most picturesque parks in the state, especially because of the beautiful setting of the Black River Gorge. This section of the river is trout heaven, with plenty of nice holding water and stretches of riffles and runs. The river is heavily stocked at no less than five locations by Fish and Game. Fly casters and spin casters alike ply the river's deep, rocky pools and undercut banks for brook trout, browns, and rainbows, often with fine results. Trout rising to fly or bait may be a stocked trout or a native fish that has migrated down one of the tributary steams. (Both Trout Brook and Rinehart Brook are left unstocked to preserve natural spawning habitats.)

The state's stocking program is augmented by the Black River Fish and Game Club, which regularly stocks trout in a posted stretch below the park. Club stockings include larger than average stocked browns and rainbows, some of which are caught at points both above and below the private club-members-only access points.

Downstream, the river undergoes a number of changes, from a change of name (it becomes the Lamington River south of Pottersville) to change of personality. Instead of the dark, deep trout stream, it becomes the kind of slower, shallower water beloved by bass and other warm-water fish. Of these, the smallmouth are the most aggressive and exciting; indeed, some believe this stretch of the Black River to offer the best smallmouth fishing in the state. The trick is to get access, since much of the land along the lower Black is privately held. However, this keeps the fishing pressure light and every hole and cut bank to the river's junction with the North Branch of the Raritan River seems to hold smallmouth.

Tip: Fish shiners on lightly weighted lines in the lower stretch of the river and remember to take plenty of bait, especially for early morning and late afternoon fishing in the spring, when the smallmouth fishing can be nothing short of extraordinary.

27 Rahway River

Directions: Take exits 136, 137, 138, or 140 off Garden State Parkway to river crossings and/or parks.

The Rahway River is the prototypical central New Jersey waterway, flowing through diverse communities and environments, starting as the outflow of sweetwater ponds and ending at its appointment with the sea. Despite significant commercial and residential development along its banks in the many communities through which it flows, the Rahway has sufficient habitat and dissolved oxygen for heavy trout stocking and excellent fishing. Moreover, it is attractive enough to be cherished by those along its course who appreciate the beauty of rivers.

The west branch of the Rahway begins in the wetlands above Millburn and just below Cable Lake and the Orange Reservoir. As Don Madson of Sportsmen's Outfitters points out, here it is a meandering field-bordered creek containing occasional pools large enough to harbor trout. The east branch of the river begins in the town of Orange. Channeled and controlled, this branch makes its way through South Orange and Maplewood. Near the confluence of Routes 78, 24, and 124, the branches join to form the main body of the Rahway River in a marsh outside of Union. The river thereupon grows in size and authority as it is fed by a number of brooks and by feeder tributaries from Shackamaxon Lake, Ash Brook Swamp, Robinson's Branch Reservoir, and Milton Lake in the areas of Clark and Rahway.

The fishing along the river is often quite excellent and always exceptionally comfortable. County and municipal governments have created parks virtually the length of the river, often including earthen berms for flood control, which are just the place for easy fishing.

Brown, rainbow, and brook trout are stocked in the river in significant numbers—usually more than 6,100 each year. Smallmouth and largemouth bass, some of them quite large, reproduce and abound in the river, as do large populations of channel catfish, bullheads, and sunfish. Striped bass and eels also surge up the river from Raritan Bay.

Big and exceedingly powerful carp are another great feature of the river, especially from Cranford down to Mountainside. RB's greatly missed fishing partner, the late Dr. Morris Weiner of Rahway, often commented on the "grass is greener" syndrome as he drove over the carp-filled river from his home in Rahway to go carp fishing 60 miles away in the Delaware River. (In fairness, however, it should be noted that the Delaware seldom disappointed him.)

> *Tip: Bait for carp with five or six kernels of canned corn threaded on a #6 baitholder hook. Use light (6-pound or 8-pound) line weighted with a sliding barrel sinker. Stop the sinker about a foot above the hook with a small split shot rather than a swivel or other contrivance, since the split shot will slide down the line rather than break if a fish is on and you happen to get stuck in a rock or other structure.*

The Rahway enters the Raritan Bay in the stretch separating Staten Island from New Jersey. Here this little chameleon of a river changes again, not for the first time but finally the last, as it finishes its trip from sweetwater to brackish to salt. As we have documented elsewhere, the bay provides wonderful fishing for a wide variety of saltwater sport fish, particularly stripers and blues. Consequently, anglers are treated to one last opportunity by this urban river finally gone down to the sea.

31 Millstone River

Directions: Access and good parking off Route 27 south of Princeton in Mercer County at the Carnegie Lake dam. Also at Colonial Park in Somerset off Route 514 (Amwell Road) to Metlars Road into the park.

The Millstone River is one of the few streams in the country that flows to the north, although it originally flowed in the opposite direction. According to geologists, its original source was a huge spring in the Sourland Mountains from which it flowed south to empty into the Delaware River. During the last ice age, however, the normal course of the river was reversed, a feat made possible by its virtually flat river valley. Now it emerges from two "new" sources: a small woodland pond near Bairdsville and a swampy region near Hightstown. They flow together to produce one of the most picturesque and "fishable" little rivers in the state.

From the juncture of its branches the river flows only a few miles to the point where it passes under the venerable Delaware and Raritan

Canal—carried over the Millstone by an aqueduct. The river then becomes Carnegie Lake, which is perhaps best known by passing motorists as the site where Princeton University's rowing teams race their long sculls. The river resumes its form below the Carnegie Lake dam and runs "cheek by jowl" alongside the Delaware and Raritan Canal—often separated only by a 10-foot towpath—from Princeton to Bound Brook, where the Millstone empties into the Raritan.

One of the most productive and accessible stretches of the river begins just below Carnegie Lake south of Princeton. With a bit of effort, a car-topped boat can be launched here. A drift downriver reveals many wonderful fishing sites, especially long stretches of lily pads and "structure" in the water such as fallen trees. We have fished the Millstone many times with fine results, often catching large numbers of crappie schooling around partially submerged trees.

Tip: When fishing such spots in small rivers, drop your anchor well upstream and drift down to the spot on a long anchor rope. In this way, your anchor digging into the muddy bottom will not be close enough to disturb your fishing.

In addition to crappie, the Millstone is home to pickerel, smallmouth and largemouth bass, sunfish, several varieties of catfish, and surprisingly large carp, the latter most often found in the larger pools along the river's course. It also attracts a number of species of ducks, a fact not lost on an enterprising RB, who decided one cold winter day to drift the river in a canoe, "jump shooting" ducks and stopping occasionally to cast live bait for bass and crappie. In the midst of all this sporting activity, he managed to turn the canoe over, losing his favorite Shakespeare spinning outfit and a good deal of dignity. A week later, ML returned to the spot, having calculated current, drift rate, and other probability formulas. After a few deft casts and retrieves with a weighted treble hook, he snagged the spinning outfit's line about 150 feet downstream of where it entered the water, retrieved it, and later returned it to a grateful (if still embarrassed) fishing partner.

Two decades ago, an industrial accident caused a fish kill of major proportions in the river. We remember standing on its banks and reflecting with sadness upon the thousands of dead fish lining both banks. Fortunately, the pollution was stopped and the river cleansed itself. Today the Millstone runs sweet and clean, its fish stocks have long since regenerated, and once again it is a river of choice for New Jersey anglers.

46 Delaware and Raritan Canal

Directions: For access to the mouth of the canal, follow Route 202 south for 26 miles to Route 29 north in Hunterdon County; go north for 6 miles to the Bull's Island Park entrance. For Stockton area, access is available directly from Route 29. Access at Kingston (in Mercer County) is available directly from Route 27 and in Somerset County directly from Easton Avenue (off Route 287).

The Delaware and Raritan Canal was once considered a critically important—indeed, the most important—artificial waterway in New Jersey. Steeped in history, the 60-mile canal was a great engineering feat when it opened in 1834, having been dug largely by strong-backed immigrant laborers with mules. Mule-drawn barges operated 24 hours a day, carrying more than 2.5 million tons of cargo in 1896, its peak year, between the Raritan and Delaware Rivers.

The last echoes of muleskinners' shouts died away with the last commercial uses of the canal, soon after the end of the Great Depression. What remains is 60 miles of clean, tree-shadowed water in the feeder and the main canal and wonderful fishing opportunities at points from the feeder mouth on the Delaware to its outfall at the Raritan River in New Brunswick.

We have fished virtually the entire length of the canal, often with excellent results. The mouth of the canal at Raven Rock is an especially natural spot for catfish and carp. Indeed, the bit of land formed on the upriver corner of canal and river is known to us as "10 – 0," a name which derived from an inglorious experience when ML landed ten carp fishing from that spot to absolutely none for RB. (In fairness, this was in the beginning of the latter's carp-fishing experience, and he had neither the proper gear nor the right attitude for success.)

Smallmouth and largemouth bass also swirl into the canal from the river at this point, as do swarms of sunfish, suckers, and the occasional eel. Fishing here with live bait, especially crayfish and large shiners, pays dividends, as does working the usual catfish and carp baits on bottom. In addition, the state stocks the canal quite generously with trout from Raven Rock down to areas well below Lambertville.

Regarding access, the canal is paralleled by New Jersey Route 29 from Bull's Island State Park at Raven Rock to a point below Lambertville. Along this long and heavily stocked stretch, access is governed primarily by finding a safe and legal place to park. (One such spot, located just opposite a stone quarry below Stockton, has yielded fine catches of stocked rainbows and browns for both authors and our children, when they were small.) This stretch also regularly produces carp in April and May.

Tip: In spring, carp prefer night crawlers to corn or prepared carp baits, for reasons known only to them. Consequently, many trout anglers dunking worms are astonished to have their spinning tackle all but ripped out of their hands by the violent strike and run of big carp. Try fishing with whole night crawlers on bottom for early season carp.

The canal parallels the Millstone River as it descends south near Princeton. The old mule towpath along this stretch provides excellent fishing access for largemouth bass, catfish, carp, and sunnies. Near New Brunswick, the canal borders and finally enters the Raritan River. It was several miles above New Brunswick that ML and RB first fished the canal together, in an incident remembered as the "great roach contest." We had discovered a wildly hungry concentration of roach (a.k.a. golden shiners) and were catching them almost as fast as we could get small hooks baited with bits of dough into the water. As often happens with us, a contest developed—to see who could catch and release twenty-one fish first. RB was catching fish as fast as he could hook, unhook, bait, and cast, but to his puzzlement, was still losing the contest. Only after ML had reached twenty-one (four fish ahead of RB) did he demonstrate his method: pinching and forming two baits at a time and concealing the second between thumb and forefinger, the quicker to get back into action.

The canal holds superb opportunities for all anglers, from novices to experts, and many pleasant memories for the authors. The entire 60-mile canal is protected as a state park, courtesy of a bill sponsored by then State Senator Ray Bateman in the 1970s. We all owe him a debt of gratitude.

62 Metedeconk River, North Branch

Directions: Take the Garden State Parkway to exit 91 in Monmouth County, then follow County Line Road to local exits, including Kent Road and Brook Road to North Branch.

The Metedeconk is a small river of large surprises. One surprise is the temperature of the water; shockingly cold while other South Jersey rivers flow tepid to warm in summer. Another is its color: crystal clear rather than the amber waters more typical of the region. Still another is the shallow layer of quicksand beneath stretches of the streambed; not dangerous because of the hard layer of sand and gravel beneath, but surprising nonetheless in a pristine little river in the Northeast.

Because of the Metedeconk's cold water, trout hold over in the stream even in the hottest summers. The skilled trout fisherman finds excellent trout fishing, especially in long stretches of the river's north branch.

This branch, although small, features the widest variety of classic trout habitats, ranging from fast-flowing riffles and long, shallow stretches to deep pools, undercut banks, and eddies that prove almost irresistible to fly casters. The upper river is regularly stocked by the state, weekly in spring and again in late fall. Consequently a good deal of the trout action is in fact "put and take" fishing. However, because the water is cold with plenty of oxygen throughout the year, many of these stocked trout hold over, often growing to fine size.

Fly fishing is complicated by overhanging trees on both sides of this lovely stream. Consequently most anglers here rely on light spinning tackle and a variety of baits, including fathead minnows, salmon eggs, garden hackle, baby night crawlers, and kernel corn.

Regarding artificials, our favorite here is the small #1 or #2 silver Mepps spinner used with ultralite spinning tackle and 4-pound test line. The short ultralite rod can be cast (more easily than flyrod or more conventional one-hand spinning rod) under overhanging trees with side-arm or underhand flips.

> *Tip: A short walk through the woods will take you to pristine pools and riffles that lazier anglers miss. However, it is important to look for posted signs, since some property along the upper Metedeconk is privately owned.*

69 Rancocas Creek

Directions: Take U.S. Route 295 to Route 541 (exit 44) in Burlington County into Rancocas State Park in Hanesport or Route 130 just east of Bridgeboro.

Rancocas Creek is actually several creeks in one, each flowing its quiet course though lightly populated Burlington County. Ultimately, the branches join to form the main branch of the Rancocas, which swells to more than 200 feet in width as it nears its outfall to the Delaware.

The north branch of the Rancocas is fed by Smithville Lake and Mirror Lake in Browns Mills. The south branch begins at Mill Pond in Vincentown. The southwest branch springs up in the vicinity of Route 70 between Chairville and Medford.

Each of the branches has its own attractions. However, our focus is upon the main and southwest branches.

The main branch is a fine muskellunge fishery. As Hugh Carberry of New Jersey Fish and Game explains, the state began stocking the creek with tiger muskies in 1989 and by 1996, a total of 20,349 tigers had been released in the creek. The stocking experiment has been very successful. Indeed, the 29-pound state-record tiger muskie caught in the Delaware may very well have come from these stockings.

In addition to muskies, the river has a fine run of spawning striped bass every spring, plus huge numbers of channel catfish that have entered from the Delaware River. Perhaps the biggest attraction for the cane pole and bobber set are the big schools of crappie and white perch found up the main stem of the Rancocas and well up several branches. In addition, big carp that would certainly cause both authors to dive for spinning tackle and cans of corn roll in the shallows on the main stem. Finally, a healthy population of suckers provide first-rate relief from cabin fever in February, which most other freshwater species are reluctant to feed.

Finally, the southwest branch of the Rancocas is stocked with trout five times each spring. In the most recent spring stocking, more than 1,100 trout were released in this shortest branch of the creek.

The average depth of the north and southwest branches is 6 to 7 feet when the water is low, and the main stem is only 2 or 3 feet deeper. At high water, the creek climbs an average of 50 percent.

Tip: Carberry tells us that if you are after tiger muskies, fish low water. You'll have more shoreline to work and the "tigers" will have fewer places to hide.

Several boat ramps serve the main body, including one at Clark's Landing Marina in Delran, not far from the place the creek enters the Delaware. Another is at Randel (named for RANcocas and DELaware) Marina in Delanco. A public boat ramp is located where the branches join in Hanesport (behind a water-treatment plant).

Stripers, tigers, fat carp, and large channel catfish make Rancocas Creek a place to try.

Fishing

Preserves

Fishing preserves—so-called "pay ponds"—provide many of the pleasures of successful fishing with little if any inconvenience. These lakes are regularly stocked with trout, bass, and other desirable freshwater fish and they fill an important niche in the hierarchy of angling opportunities. For the neophyte angler, the parent taking a child fishing, even the more experienced angler who is short on time, these well-stocked lakes provide an almost sure-fire experience, with fish and fun to show for the effort. The following are three of the best such places, each in a different region of the state.

12 Drake Lake

Directions: Take Route 94 north to Route 616 (Newton-Sparta Road) near Newton in Sussex County. Follow 616 east through Newton to Drake Lake.

Drake Lake is also called "Go Fish," and it is difficult to imagine a more appropriate second name. The beautiful setting of this 9.5-acre natural lake could hardly be more conducive to fishing. Anglers have the option of fishing from shore or rental boats for a surprising variety of stocked fish, including hybrid bass, channel catfish, largemouth bass, pickerel, crappie, and sunfish. It is heavily stocked with trout, which feed especially well from the time the lake opens to fishing on March 15 through mid-June. The lake is licensed by the state as a private fishing preserve. Consequently no freshwater fishing license is required. However, clearly posted rules must be followed. Most recently, these rules stated that all largemouth bass must be released, and that all trout, hybrid striped bass, and catfish must be kept (and paid for). Any other fish may be kept or released at the angler's option. Common sense dictates that if a fish swallows the hook and is injured, that fish should be kept and paid for. More than incidentally, charges for fish are by the pound.

> Tip: *If fishing live bait or throwing artificials, keep your drag open against the possibility of a hybrid bass strike. Hybrids are real line-busters, especially on the strike, which is so sudden and powerful that it almost seems jet-propelled (hence our nickname "rocket" for these wonderful game fish).*

The lake has some exceptionally large pickerel, which vie with the bass for live bait, plus large populations of crappie, perch, bluegills, and common sunfish. All are fun to catch, especially for the kids, and make a fine meal.

"Go Fish" has its own tackle store, where tackle and bait may be bought, and where fishing outfits can be rented by the casual angler. Boat rentals are also available, with or without electric motors, by the hour or by the half- or full day. Indeed, everything favoring a pleasurable fishing experience—for a memorable family outing or for the more serious angler—is provided here.

48 Molder's Ponds

Directions: Take Route 18 south from New Brunswick to Englishtown Road. Follow this road about six miles to Molder's.

Several years ago, ML joined a group of outdoor writers at these heavily stocked ponds and all concerned had a ball. The fact that outdoor writers—all accustomed to fishing the finest and most pristine lakes and rivers in the state—convened at pay ponds tells you everything about the quality of fishing they afford.

Molder's is actually seven ponds in all, with the three largest ranging from 2 to 3 acres and with depths up to 28 feet, providing fine conditions for fish habitat and growth.

Immediately accessible from a number of central Jersey cities, Molder's is thought to be the only "pay pond" in the state stocked with walleyes. Large numbers of trout are stocked in the spring and fall as well. (Significantly, trout are the only fish that must be kept and paid for when caught.) Heavyweight channel catfish and plenty of common bullheads also abound in the ponds, along with largemouth bass, bluegills, eels, and hybrid bass. Any of these species may be released or taken home if paid for.

Tip: Talk about what you will do with fish before your kids begin to catch them. Advance agreement as to what you will or will not take home means no disappointments on an otherwise fun-filled day.

Hooks must be barbless, which allows fish to be unhooked and returned to the water with the least amount of trauma. If a fish swallows a hook, it is a good idea to cut the line as short as possible without excessive handling of the fish or trying to retrieve the hook.

No boats are allowed on any of the seven ponds. However, the banks are wide and for the most part free of weeds, snags, or trees, which makes for very comfortable fishing. An entry fee is charged, but this fee usually turns out to be a bargain—especially when bringing a child fishing.

95 Bauer's Fishing Preserve

Directions: Take the Garden State Parkway to Route 9, Seaville, in Cape May County.

Big, powerful fish are the rule rather than the exception at Bauer's Preserve, distinguishing it from many "pay lakes" which feature swarms of smaller panfish to delight the youngsters.

The featured fish at Bauer's is hybrid striped bass, which also happens to be our absolute favorite game fish. Stocked as 12-inch fish a few years ago, the hybrids have grown at an incredible rate in this 2-acre, spring-fed pond. Ten-pound hybrids were being caught as early as 1994 and the existing state record of 10 pounds 14 ounces has been unofficially broken here three times. Since the lake maintains a strict catch-and-release policy, all of the trophy hybrids have been returned to the water to fight again another day. Given the growth rate of these fish, it is thought that some hybrids may have reached 14 pounds in size—an unimaginable "beast" to those of us who have battled 5- and 6-pounders in other places.

Brook, brown, rainbow, and tiger trout are stocked in early spring and again in October, ranging in size from 13 to 18 inches, plus a few heavy-weight browns up to 10 pounds. Largemouth bass also are stocked to augment a naturally reproducing population, and bucketmouths of 4 to 8 pounds have been released since 1995. All largemouths must be released.

Channel catfish are another featured fish in the lake. They are aggressively stocked in spring, summer, and early fall. The average fish weighs between 4 and 5 pounds when stocked, and like the hybrids, their growth rate is excellent.

Tip: The channels show a marked preference for chicken livers, but occasionally a hybrid will pick one up as well. In either case you're in for a good scrap.

Large populations of crappie and bluegills—some weighing as much as a pound—help make this a wonderful experience for youngsters. With the panfish as well as the game fish, large fish predominate.

The rules at Bauer's are clear. Anglers pay a fee to fish. No license or trout stamp is required. Hybrids and largemouths must be released, but all other species may be kept (for a separate charge).

Rods and reels may be rented and bait purchased right at the lake.

Saltwater Fishing

Saltwater
Rivers
and Bays

Unimaginably rich in nutrients and marine life, saltwater rivers and bays provide a virtual paradise of opportunities for anglers, crabbers, and clammers of all ages and levels of skill. Although all of these sheltered waters share certain characteristics, they are differentiated by their shape and structure, their size and depth, their sources and their salinity. Fished from boat or shore, pier or jetty, the following are among the best of a wide spectrum of wonderful choices.

33 Raritan Bay

Directions: Route 287 to Route 35, which touches the bay at Laurence Harbor. Take Route 36 east along the Bay, then local streets to the water from Keyport to Monmouth Beach.

The outfall of the Raritan River meets the sea at Perth Amboy, near the tip of Staten Island. The bay is an extremely large body of water that curls all the way down to the Earle Ammunition Pier on the New Jersey side and mingles with the waters of Lower New York Bay to the north and east.

Raritan Bay is a truly wonderful resource for anglers, especially those in search of striped bass, bluefish, flounder, and fluke. We recall that a sturgeon was hooked in the bay some years ago.

April through June are among the best months to fish the bay, especially for flounder and schoolie bass. Particular hot spots are Princess Bay, Seguine Point (near Lemon Creek), and the Conference House.

A favorite shore spot is a stone barge off Highland Boulevard, where anglers plug for bass, blues, and weakfish as the season progresses. The bogs at Great Kills Park is another first-rate spot to throw swimming plugs and bucktails with Mr. Twister tails.

Many party boats and private craft fish inside in the bay, especially when the ocean is too cold for good fluke fishing. Anglers aboard these boats often bring excellent catches of flounder and fluke as well as warm memories back to the docks. (It is important to remember that New York and New Jersey laws are often quite different, especially as they pertain to flounder.)

Fluke action improves in the bay with the warmer weather in May. The Bug Light plus Hoffman and Swinburne Islands produce early; then, as the fluke swim to deeper waters, the action is often excellent near the western end of the Raritan Reach.

Surf casters begin to score on striped bass in late March and early April, especially at Morgan, Cliffwood, Pebble (Union), and Port Monmouth beaches.

Tip: Go with blood, tape, or sandworms on sliding fish-finder pyramid sinker rigs with a 3-foot leader on either side of high tide, preferably very early in the morning or on heavily overcast days. Hang on to your rod, because a big "cow" could drag it into the bay in a heartbeat.

Schoolie stripers appear in greater numbers later in April. Anglers score well under the Victory Bridge or the Spy House at Port Monmouth with sandworms fished just off bottom.

The bay continues to produce right through summer. Incoming tides produce super fishing for fluke, blues, and weakfish in the hottest part of the summer. Fishing sandworms along the drop-offs in Reach Channel, especially between buoys 6 and 12, is especially deadly on sea trout in summer. (One rule of thumb: "When azaleas abound, weakfish are around.")

Figure 12 ■ Captain Scott Vigar with typical Raritan Bay area weakfish.
(Photo: Captain Joe Kasper.)

As the water begins to cool in the fall, blackfishing heats up, with particularly good catches recorded at buoy 44 and at the Staten Island Monastery. Huge bluefish go for small peanut-sized live bunker at this time, and striped bass are taken on live eels, bunker, herring, or plugs, especially around dark.

The best-kept secret in the bay is flounder in fall. The spring spots we mentioned, plus the main channel at Crab Island off South River and Morgan Creek at the Parkway Bridge, are hot spots for exceptionally large, hard-fighting, and tasty "flatties."

Boats may be rented at the marina under the Morgan Bridge, and boat ramps for launching private craft are available at the above-mentioned marina, at the end of the Municipal Fishing Pier in Keyport, at the South Amboy Boat Club, in Sayreville, and in Seawarren on Route 35.

Shore fishing spots abound around this massive bay. Perhaps most popular is the Keansburg Pier, located at the end of Main Street in town, and the Keyport Pier, both of which charge a fee but produce fine fishing action for bluefish and weakfish as well as flounder, fluke, porgies, and crabs. Other first-rate fishing spots are the Laurence Harbor jetty, the Perth Amboy Pier, and the bridge at South Amboy. Surf casters score well from the sand at South Amboy, Morgan, Cliffwood, and Laurence Harbor, among many other locations.

34 South River

Directions: From exit 9 on the N.J. Turnpike, take Route 18 South less than 1 mile to Edgeboro/Old Bridge Turnpike exit. Take jughandle toward town of South River. Turn left on Burton Ave., to end. Left on Williams St. to parking in Varga Park and shoreline.

South River is a flowing set of contradictions: a little river with big and enormously varied angling possibilities, a river that is brackish to salty or fresh from one tide change to the next, a river flowing through highly developed areas but still providing quaint pleasures of wilderness fishing.

Many of these contradictions have to do with size. South River is only 8 miles long, from the pond where it begins just west of Richland on Route

40 to its confluence with the Raritan. Moreover, the river is almost entirely tidal, so at high tide, saltwater fish flood up the Raritan and into South River, bringing striped bass and great schools of white perch miles upriver. Being tidal, freshwater species feed with great enthusiasm on the outgoing tide, when fresh water again predominates in the river. ML remembers spotting a huge school of yellow perch from a bridge in the town of South River many years ago, fish that would not take bait, presumably because they saw him. Now white perch predominate in the river and are taken in numbers on live grass shrimp.

In the spring, thousands of herring run up South River to spawn. Weighing up to a pound, these fish show a marked preference for bare gold hooks, which they hit with great enthusiasm.

Tip: For fluke fishing, herring fillets (with skin on) are outstanding bait. A mess of herring taken in the river, filleted and packed skin to skin in small packages of aluminum foil and frozen, will pay big dividends in flatties later in the season.

Angling from shore is not only possible along the South River, it is actually the preferred method by a majority who fish here. Along its upper reaches, at about the point where Route 50 is intersected by Route 40, plenty of shore access is available, and the fish of choice is pickerel. In fact, the pickerel are so plentiful here that one angler is said to have caught and released more than 700 in one short stretch of the river in one recent season. Minnows and artificials that emulate minnows produce best. Healthy populations of carp and catfish also inhabit the river.

What makes South River so very interesting is the quality of its fresh- and saltwater fishing despite the development along its shores. Aficionados of a different sport know the town of South River as the birthplace of such great NFL athletes as Joe Theisman, Kenny Jackson, and Drew Pearson. Even more interesting to us, however, is the river that gave the town its name and its infinite angling opportunities.

36 Sandy Hook Bay

Directions: Take Garden State Parkway to exit 117; then take Route 36 to Atlantic Highlands. Follow the coast along the Bay.

The Sandy Hook region straddles coastal sections of Middlesex and Monmouth Counties and provides many classic in-shore fishing spots. Our personal favorite is Sandy Hook Bay, where action starts to heat up in early spring and never cools down until the coldest winter months.

Early in April, anglers baiting with skimmer clam, herring, or mackerel strips start to catch ling in significant numbers just inside the bay at buoy 1. Later in April, hungry flounder in large numbers leave the rivers and wintering holes in the bay and start to feed voraciously on offerings of sandworms, bloodworms, and clams. Special flounder hot spots include the Pump House near the Leonardo Flats and Plum Island at the mouth of the Shrewsbury River. Another dynamite spot for huge April flounder is the Oil Dock between the Navy Pier and the Atlantic Highlands Marina, which can be located by on-shore oil tanks as landmarks.

Striped bass begin to arrive in Sandy Hook Bay about the time the magnolia trees start to blossom, usually in mid-April. Before the month is out, some fluke also appear in the bay.

> Tip: By mid-May, clouds of diving gulls announce the arrival of marauding schools of bluefish, often in the warmer waters close to shore. Get to those areas quickly and throw metal (an Ava-type A-27 jig without teaser works especially well). Make sure your drag works well though, because these skinny blues can pop your line in a heartbeat.

Memorial Day is a reliable time to begin drifting the bay for fluke, and in June, fluke in large numbers start to get serious, especially across the bay at Officers' Row and around the Pound Nets. Trolling big spoons and umbrella rigs during the day (and plugs at night), especially out near the end of the bay at the rips, are good ways to catch stripers.

Super fluke catches begin with the tee-shirt weather of July, especially along the drop-offs along Pound Nets and Officers' Row. Standard baits are recommended, but anglers also have luck here with a white 3-inch

Figure 13 ■ Josh Kasper holding an Earle Ammo Pier area fluke at Sandy Hook Bay.
(Photo: Captain Joe Kasper.)

Mister Twister, Hootchie squid in white and/or chartreuse with killy. We've also had luck here with a #1 nonweighted white bucktail and small killy, held 18 inches above the sinker on a 6-inch leader.

More great fluke fishing is found at the drop-offs right in front of the Navy Pier on the red 2 and 4 channel and Terminal Channel buoys 2 and 4 red or 1 and 3 green.

Summertime also signals showings of stripers and weakfish in Sandy Hook Bay, the former from Horseshoe Cove to the Bug Light; the latter around the Navy Pier and near Leonardo Marina. Effective offerings for weakfish generally involve sandworms, either plain or "à la mode" (that is, combined with a strawberry jelly worm combo or other concoction). The bay also is loaded with snapper blues all summer.

In October the striped bass is king of the bay, with trophy stripers taken while trolling bunker spoons near the Pound Nets, the Bug Light, and the channel leading into the Shrewsbury. Headboats score well on linesides in the evening on incoming tides, using whole sandworms or live eels. Porgies, small sea bass, and keeper weakfish are regularly taken in the bay as well at the top of both tides, especially at Spermacetti and Horseshoe Coves. Fat, hungry flounder also are active from October through December, rounding out the calendar year in this great fishing area.

Rental boats and motors are available at Atlantic Highlands. The state has a small boat ramp in the marina at Leonardo, which is reached by turning east on Route 36 and working your way through a series of odd turns. Larger launch facilities are available at the Atlantic Highlands Marina. Both ramps require fees.

43 Navesink River

Directions: Route 35 south to Navesink River Road below Middletown in Monmouth County.

The Navesink River is a rich feeding ground for marine life and a magnet to anglers and crabbers alike. The river is brackish from the outfall of the Swimming River Reservoir in Shrewsbury Township to the river's confluence with the Shrewsbury below Rumson and its junction with Sandy Hook Bay.

Like other saltwater rivers, the Navesink is active in early spring with schools of flounder just up from winter hibernation. For about two weeks when water conditions are just right, flatties feed well before heading out to open waters. April through June finds first-rate action on bluefish, with 6- to 8-pounders chasing live bait and artificials. Pete Pawlikowski, who operates the Oceanic Marina in Rumson, is one of a number of first-rate local anglers who watch the Navesink for diving, screaming birds. When he sees them, he is off like a shot for his dock to cast Hopkins and other metal for super action on blues.

Fluke appear in June and continue to feed right through the end of summer, especially in holes and drop-offs near the bridge and buoy 18. Summer weakfish also are taken in these areas, especially on early tides with shedder crab and doodlebugs.

Tip: When fishing for weaks, lift your rig (baited with sandworm) and let it drop. Weakfish tend to hit as your bait falls.

Be on guard for what the locals call "Sally-Growlers" and what others call "Uglies," or "Oyster Crackers." These fish, which are very much like headfish, toadfish, and monkfish, are all head and appetite. A baited hook on bottom, especially on a slack tide, often means a considerably less than desirable fish on your deeply swallowed hook. (We recommend cutting your line rather than trying to remove the hook, since these sharp-toothed creatures can wreak havoc on a finger.)

August often brings great kingfishing under the Oceanic Bridge, with sandworms the bait of choice. Snapper blues are everywhere in summer and so numerous that anglers intent on fluke have to remove their white attractors (squid, etc.) and fish an unadorned live killy to keep bait in the water.

Crabbing is at its absolute peak during the warm months of summer. Traps and drop lines both produce well, as does dip netting at the pilings on an outgoing low tide. (Remember that females with eggs showing must be released.)

Good access is available at Marine Park in Red Bank, where anglers—especially young ones—enjoy crabbing and snapper fishing. Shoreline access is available across the river from Oceanic Marina on the northeast side of the bridge. (Fish here on a high tide, since there is a four-foot tide drop.) A third spot—especially for snappers, fluke, blueclaw crabs, and flounder—is a very comfortable pier at the end of Fair Haven Road in Fair Haven. A rule of thumb here is fish on the high tide and crab on the low.

There are boat-rental facilities at Oceanic Marina and back up the river at Red Bank. Two boat ramps service the Navesink: one at the west side of the Red Bank Bridge and a yearly permit ramp behind Borough Hall in Rumson.

49 Shrewsbury River

Directions: Take Garden State Parkway to exit 117; then take Route 36 to Sea Bright. Follow local streets to river access above and below Monmouth Beach.

Unlike many other New Jersey rivers which begin in some distant sweet-water pond or upland marsh, the Shrewsbury is entirely a saltwater river, beginning at a point just inside Sandy Hook Bay. There are no freshwater/saltwater confusions here: the bountiful catches of fish are all saltwater species. Most importantly, these catches are often superb, whether fishing from a boat or from many excellent spots on shore.

The best boat fishing is found in several clearly identifiable spots. The best are associated with major bridges, beginning with the Quay (pronounced "key"), which is located just beyond the Highlands Bridge coming from Sandy Hook Bay. Quay fishing for flounder, striped bass, and fluke is often outstanding, depending upon the season and the tide. Slack water is best from an hour before to an hour after slack water. Flounder action is excellent in early spring and gets better with passing days. Flounder action also is good upstream toward the Rumson/Sea Bright

Bridge and down toward the mouth of Sandy Hook Bay. Another favorite spot is Graveley's Point where the Shrewsbury and the Navesink meet at Buoy 32.

Many striped bass are taken in late spring under the Highlands Bridge, just north of the Quay, at the Rumson/Sea Bright Bridge and the old Turnstyle Bridge.

Tip: When fishing for early-season stripers, rig as many as three or four sand-worms on a hook, on a dropper rig with no float. Bounce bottom through the channels and bridge abutments while drifting, preferably in early morning.

Blackfish also can be caught under the Highlands Bridge on fiddler or green crabs at slack tide. However, the focus in summer tends to be on fluke and bluefish, and the blacks are most often ignored. Legal-sized weakfish, generally in the 15- to 20-inch range, feed well along the channel edges in warmer weather up and down the river.

In summer, crab aficionados do exceptionally well in the river on slow tides. Crab traps as well as drop lines often produce bushel baskets of delicious blueclaws, with August the best month of all.

Snapper blues flood the entire river in summer, and massive fluke follow. As the weather cools, striped bass begin to feed seriously on live bunkers and eels under the bridges, especially on an outgoing morning tide.

Regarding boating, there is a boat-rental facility on the river at Highlands and a small, free boat ramp located at the end of Atlantic Avenue in Long Branch.

No less important, there are plenty of fine spots to fish from shore. One is the bulkhead at Sea Drift Avenue and Washington Street in Highlands (although you cannot park and start before 9 a.m.). A park in town at Bay Avenue produces good flounder action from shore. (Use worm and two hooks, one below the sinker and another 12 inches above it on a 6-inch leader. You'll catch more flounder on the high hook.) Early-season blues are caught at the bulkhead in downtown Sea Bright. On the east side of the river, walk from the parking lot on Sandy Hook across to the river side and fish near Plum Island. Perhaps the hottest shoreline spot is the former bridge abutment site on the north side of the Highlands Bridge.

58 Shark River

Directions: Garden State Parkway to exit 100B; take Route 33 to Route 35 to Belmar Marina.

The Shark River is one of the smallest to flow into the sea along coastal New Jersey but one of the largest in terms of fishing excitement. Three areas of the river are especially productive: the Shark River Hills, the North Channel, and the area of the Belmar Marina.

The Hills are reached from Route 33 east, turning on Route 17 west, Sylvania Avenue, to the river. This area of the river is first-rate for winter flounder, especially on higher tides. (ML has fished for founder here no less than 100 times, almost always with good catches.) In June fluke start to show up, and casting and retrieving from boats anchored on drop-offs is recommended, since the area is too limited for effective drifting. As the weather gets hotter, so does the fishing for fluke, plus astonishing numbers of snapper blues to delight kids and grownups alike. In October kingfish and blowfish also appear in the clean water in the back of the river, and the best flounder fishing of the year takes place in November and early December in the Hills, especially sandwiched around high tides.

The North Channel is best accessed from the area around the Route 35 bridge. It is a fine springtime fishing spot for flounder, especially for the enterprising angler who knows how to chum.

> Tip: *March is good, April is better, and May is best as feeding flounder get ready to head out to sea. The fishing is often best east of the Main Street, Route 71 bridge, which also is a good place to fish for fluke from July through September.*

Figure 14 ■ Two fat flounder caught at the Shark River Bulkhead.
(Photo: Captain Joe Kasper.)

Drifting between the Route 71 and Ocean Avenue bridges is highly recommended, especially on incoming to high tides. Around September a phenomenon which expert Lou Rodia used to call "potpourri" takes place: that is, you never know what kind of fish you can catch here on an incoming tide. ML took eleven different kinds of fish just east of the Route 71 bridge one day in late September. They ranged from flounder to fluke, kingfish to porgy, snapper to jack.

Over the Route 35 bridge and around Belmar Marina proper, the area from the southern tip of Maclearie Park back to the headboat dock is very good for cold-water flounder. The fishing gets better in May and June, and fluke feed ravenously on killie and squid in July, which some feel is the best month in the river. Crabbers are very successful in this area of the river in August. In September good-sized weakfish are caught at night on sandworms, especially at high tide. Flounder fishing gets hot as the water starts to cool, making the river far and away the most popular in New Jersey for late-season flounder fishing.

The flounder fishing is particularly productive for the well-equipped fisherman, which in ML's case includes two empty bushel baskets, a hard garden rake and a softball bat, along with the more usual rods, reels, tackle box, etc. He utilizes this odd collection of gear in the following fashion: First, he stops at a special spot on the way to the river and fills one of the baskets with small stones. Second, he visits the underside of a special floating dock (at low tide) and, with the garden rake, fills the second basket with black mussels, which he thoroughly smashes underfoot.

Third, he anchors at a channel edge and gets down to business. Baiting with bloodworms, sandworms, wild mussels, and bits of clam, he gets his weighted lines overboard. Then he throws occasional handfuls of stones overboard (the little puffs of sand they kick up simulate crustaceans burrowing) alternated with crushed mussels (for chum). For good measure, he throws an occasional skimmer clam aloft and smashes it downward with the softball bat, driving bits of shattered shell and meat toward the bottom among his terminal baited hooks.

For a while, fishermen in sight of this spectacle watch ML with mingled amusement and incredulity. Soon, however, they note that while they are catching one flounder, he is catching ten. That's when boats start to ease closer and closer to manic Manny.

If you need to rent a boat, there are two good facilities: one right in the Belmar Marina and the other just off Route 71 at Fifth Avenue. There are three boat ramps, each corresponding to one of the areas of the river described here. The first is in the Shark River Hills Marina, the second in the North Channel, and the third and widest (with the largest amount of parking) is just before Maclearie Park.

A boat is not an absolute necessity. In fact, shore fishing from the dock just south of Belmar Marina and the private docks at Shark River Hills plus shoreline areas at each of the bridges often produces outstanding results.

For anglers interested in variety with a salt flavor, this wonderful river is hard to beat.

63 Manasquan River

Directions: Take Route 71 or Route 35 south in Monmouth County; follow local streets to the river in Point Pleasant, Brielle, or Manasquan.

Everything that makes a river exciting for anglers can be found in the Manasquan, from its trout-filled headwaters to its broad lower tidal reaches where good fishing for saltwater species lures the angler from March to December.

The upper Manasquan is one of the most heavily stocked trout streams in New Jersey, receiving as many as 11,000 trout per spring stocking season plus several thousand more in the fall. Marked by deep pools and stretches of long, shallow riffles, the Manasquan is considered one of Central New Jersey's best and most productive trout streams.

However, it is the exceptional saltwater fishing in the tidal section of the Manasquan that we want to share with you here. This fishing features flounder, fluke, blackfish, weakfish, snappers, sea bass, sand porgies, blowfish, kingfish—a virtual potpourri of angling fun.

The trick is to find the right spot in the right season. In the coolest part of March, the best spot for flounder is the south side of the Mantoloking Bridge and what is called Gunner's Ditch just north of the bridge. An outgoing tide and sandworm are the right combination for flatties here.

In April some big striped bass move into the river, feeding freely on bloodworms and sandworms held off bottom with barrel sinker/slider float rigs. At this same time, the flounder action increases, often upriver above the Route 70 bridge. Later in the month, they are found in large numbers from the Route 35 bridge to well east of the railroad bridge. (Once east of the bridge, the activity is often wild on the south side of the river off of Gull Island, dead on the edge of the channel drop-off.)

In May blackfish come into the river to spawn, hitting clam and sandworm bait early but gradually switching preferences to fiddler and green crabs. Stripers rip into live eels, clam, and worms at the bridges at night.

In June fluke come into the river to replace the retreating flounder. A prime spot is Gunner's Ditch and up the mouth of the Metedeconk River, which joins the Manasquan. Another is the old channel in front of Clarks Landing Marina. At the same time, many smaller stripers are taken on bucktails and other artificials from the Point Pleasant Canal to the Route 35 and "train" bridges.

Crab lovers in small boats ply the river in July, working the pilings of bridges at low slack tide with dip nets, especially around the Route 70 bridge. At night, weakfish action is good, especially from the Mantoloking Bridge and the canal mouth.

> Tip: Try either a sandworm on a smaller (#1 or #2) hook or a jigged bucktail and rubber worms for weaks. Retrieve your lure erratically across tide, bouncing bottom. Incoming to high tide at night is best.

Weakfish continue as the fish of choice (but by no means the only fish) in August. Dark hours and sandworms moving slightly in the tide are a deadly combination, especially around the entrance to Cook's Creek, at Gunner's Ditch and the mouth of the Metedeconk.

In September, a dozen varieties of fish pour into the river, making it the perfect spot for kids. Bar jack, snappers, blowfish, sea bass, sand porgies, flounder, and spot (lafayette) compete for bits of supermarket shrimp, blood- and sandworms, or skimmer clams in a few areas, especially near the railroad bridge and upriver from the Route 70 bridge. Seldom weighing more than a pound, these fish make up in numbers and appetite what they lack in size. Super kingfishing is found in the drop-offs. And weakfish will go wild for a live-lined spot near the Mantoloking Bridge, at the canal mouth near the hospital, at Cook's Inlet right into north Barnegat Bay.

As the weather begins to cool, the biggest fish of the year start to feed more actively, including striped bass in October and well into November (on live eels, plugs, or bucktails). November also sees good flounder action spread deeper into the river.

There is a boat-rental facility and a boat ramp at Glimmer Glass Creek in Manasquan and additional ramps at 381 Brielle Road, in the canal on Bridge Avenue in Point Pleasant, and at the end of Bay Avenue in Point Pleasant. Good shore-fishing access is found on the Maxon Avenue Pier near the Point Pleasant Hospital. Other good shore spots include the mile of canal at Bridge Avenue in Point Pleasant, the western-side bulkhead on either side of the Route 88 bridge, and the bulkhead on either side of the Route 35 bridge.

72 Toms River

Directions: Garden State Parkway to Route 37 exit, then take local streets southeast to the river.

Toms River is another of South Jersey's sweet to brackish to saltwater streams that produce excellent fishing in every season. The upper (western) 10 miles of river are cold, clear water running through wild swamps. This is trout country, stocked by Fish and Game in spring and fall, with many trout holding over and reaching nice proportions. A stretch of the river north of the Route 571 bridge is a special conservation area where only artificials can be used and the limit is one fish per day, 15 inches or longer.

To this point, the river narrows in spots to a width of 15 feet. At South Lakewood, however, it begins to open up, and by the time you reach the area east of the Parkway toward Lavalette, the river is a major body of water. At this point, it is chock full of nice-sized white perch, caught in big numbers through the ice in the dead of winter. ML, who is always cautious about ice, caught a fine mess of "blue nose" perch through the "hard water" here, using live killies. Best baits also include grass shrimp and pieces of bloodworm.

> *Tip: Re. ice fishing, as Delaware River Park Ranger Dave Bank once told us, if you don't see other people already fishing on the ice, there's no sure way to know it will support you. Good advice.*

The perch are still around in open water in March, and the best time to catch them is afternoon when the river is running slowly.

Flounder are caught in the lower river around Island Heights and under the Pelican Island Bridge (in Barnegat Bay itself). More than one angler expecting the tentative bite of a flounder has had a worm-baited bottom hook slammed by a passing striped bass in the 12-foot depths of the river.

Striped bass become much more active in June, and they feed actively (especially on drifted sandworms near bottom) until the water becomes too warm. Small bunker can be netted in July and used to catch stripers from dusk until nightfall with a bobber and small split shot. Whole shedder crabs on long leaders also produce striper action.

Summer also brings nice weakfish action at night at Huddy Park in the town of South Toms River, and snapper blues and spot provide hours of fun there in August and September. Snappers respond well to small spoons. (Among the best shore-fishing spots on the lower river is at Huddy Park, where as much as a half mile of bulkhead exists for public fishing. To reach it, take the first right after leaving the Parkway onto Route 37 and go to the Main Street Bridge.)

No boat rentals are available in Toms River itself, but rentals can be found in Bayville at the juncture of Toms River and Barnegat Bay. Another is located off Route 9, 4 miles south of the Parkway, in Bayville. A free boat ramp can be found at "Trilco," across the Main Street Bridge in South Toms River, at the old Bus Station. From the trout fishing upstream to perch to stripers below, this river is not to be missed.

73 Forked River/Oyster Creek

Directions: Route 9 into Forked River.

What creates good fishing in various spots is often a break with the ordinary: a reef, a deep hole, a fast stretch of water, a riptide in the surf, a downed tree, or the pilings of a pier in an otherwise flat piece of bottom. Certain environmental changes also improve fishing and Forked River/Oyster Creek is a first-rate example of this phenomenon. The change here is caused by warm water discharged from the famed Oyster Creek Power Plant.

The Forked River has three branches: north, middle, and south. The southern branch brings cold water to the power plant, which discharges the warmed water into Oyster Creek just below the south branch of the river. When the plant is operating, especially in the coldest months of winter, it activates flounder that might otherwise be dormant, often providing fine catches when the "flatties" are feeding nowhere else. The action is especially good on the western side of Oyster Creek.

Tip: When fishing this particular stretch of the creek, bait with bloodworms or sandworms, and use a larger sinker since the water is fast here.

The creek discharges into Barnegat Bay, where the flounder action heats up in March. Among the best spots are three small holes south of Buoy 34 in what is called the "Whale Bone." Flounder still are feeding in April and May, but anglers tend to pay more attention to the blues and bass passing from the bay all the way up to the mouth of Oyster Creek. Both species take bloodworm or shedder crab, especially at night. Plugs and metal produce fish too.

In early summer, weakfish appear, especially around Berkley Island, and are susceptible to sandworm, shrimp, shedder crab, and bucktails.

Walt Whalen, an authority on the area, favors a half-ounce spear-headed white bucktail tipped with a small piece of shedder crab and worked slowly along the bottom. Weakfishing continues strong right through the end of summer, in the bay, at the mouth of Forked River, and at Tices Shoal. Summer also brings striped bass action in the river and around Mud Channel and the entrance of Oyster Creek. Live eels, bunker, or herring are deadly on bass. The summer is also especially productive here for crabbing aficionados.

Bass, weaks, and fluke continue to be found in this vicinity through early fall. In November, bluefish often make their way into the bay and back up the rivers, feeding voraciously and adding to the fun.

Anglers do not necessarily need boats to get in on the action, although boat rentals are available at the Bayville/Lanoka Harbor line and in Waretown. In fact, some super shore fishing is available in a park at Berkley Island, which features a fishing pier, parking, and bathroom facilities. In addition, superb fishing is available in Oyster Creek on either side of Route 9. (Anglers on the west side of the trestle must have a freshwater license.)

Another good shore spot is the bridge at Beach Boulevard in Forked River. Boat ramps are available on Lacey Road in Forked River and at Marine Plaza off Lakeside Drive East in Forked River.

76 Barnegat Bay/Long Beach Island

Directions: Take the Garden State Parkway to Route 72 east onto the island.

Long Beach Island (LBI) is a magnet for summer vacationers who flock to its fine sandy beaches to bask in the sun. However, despite the traffic and the fuss, the LBI side of Barnegat Bay provides fine opportunities for fishing and crabbing even in summer. And when cooler weather arrives and the crowds go home, anglers are queens and kings on this beautiful 13-mile island.

The coldest winter weather drives most fish out to deeper, warmer water, but in early spring, the flounder fishing heats up, especially along the Fifth Street bulkhead and areas around Harvey Cedars and High Bar Harbor.

Tip: Fish the outgoing to dead low tide with bloodworms, strips of skimmer clam, or a combination of the two for first-rate "flattie" action. And don't forget to chum.

As the water warms a bit more in April, May, and early June, the sun worshippers still have not invaded Long Beach Island in huge numbers, but blackfish and weakfish have invaded the bay, with a few fluke in the mix. The "tog," which include many big spawning females, are especially fond of feeding along the western end of the Causeway Bridge on the north side. They show a marked preference for bloodworms early on, when their jaws are too tender to enjoy their usual crab meals. Weakfish stay in the area to feed through the summer and are regularly taken on grass shrimp and shedder crab. Nighttime anglers do especially well on "weakies" when (1) fishing in illuminated areas and (2) chumming aggressively with grass shrimp. The White House, located at Post Island near the Parker Islands, is a top weakfish hot spot in late summer.

Bluefish swell the action through the summer and fall, and as the water cools, blackfish return to the area of the bridges. Often the presence of small blues will be an indication that bigger weakfish are located just below them.

Barnegat Bay also is a fine venue for crabbing. Delicious blueclaw crabs are taken in numbers into October and, in mild weather, well into November. Clamming is another productive and extremely popular activity here.

A first-rate shore-fishing spot is a large expanse of bulkhead in Ship Bottom that yields fine catches of flounder and fluke, weakfish and crabs.

LBI features quite a few ramps. To the north, there is a pay ramp at Tenth and Bayview in Barnegat Light. Further south is the Municipal Surf City Ramp at Barnegat Avenue and Division Street. There are also municipal ramps at the foot of the Duck Island bridge in Beach Haven at Ninth Street. Boat rentals are found on Bayview Avenue between Seventh and

Ninth Streets at Barnegat Light, on the north side of the causeway at Cedar Bonnet Island via Route 72 westbound, in Ship Bottom south of Route 72, and at Waverly and Twentieth in Beach Haven Gardens.

81 Atlantic City Area

Directions: Atlantic City Parkway or Route 30 directly into Atlantic City; local streets to fishing sites.

In the minds of many, Atlantic City is synonymous today with casinos, high-rise hotels, and condominiums. However, mention Atlantic City to anglers and visions of striped bass, sea trout, bluefish, blackfish, fluke, and white perch dance in their heads.

The Atlantic City area is in fact a confluence of incredibly rich grounds for all kinds of marine life. It is, after all, the point where the drainage of the entire Brigantine Wildlife Area— one of the largest, richest, and most diverse coastal wildlife refuges on the East Coast—joins the ocean. It is the thin buffer between Absecon Bay and Egg Harbor and the Atlantic. It is also the point where the Great Egg Harbor River, Middle River, and Tuckahoe River turn from brackish to salt. More to the point, it provides superb conditions for a spectacular variety of saltwater angling opportunities.

Of course, the Atlantic City fishing area is vast; consequently, knowing the best spots for various species is critical. For example, white-perch fishing is great in January and February, especially in the mouths of the Great Egg Harbor, Middle, and Tuckahoe Rivers. Grass shrimp is the bait of choice on the top and change of high tides. In March, the perch tend to move out into deeper waters, with concentrations in Great Egg Harbor Bay behind Beesley's Point at the Power Plant.

A run of fluke takes place in April. The Ship Channel between Ocean City and Somers Point is an early stopping-off point in shallower water first. Striped bass are the hot fish in April, in Great Egg Harbor River, around the Longport/Ocean City Causeway, at the Longport Rockpile, and just inside the Atlantic City Inlet. Floating Rebels and Bomber plugs score well here.

Winter flounder are caught in April at Rainbow Channel or Ship Channel, both behind Ocean City, and Lakes Bay behind Ventnor. A combination of bloodworms for bait and aggressive chumming will produce fish.

Once water temperatures exceed 50 degrees, generally in May, bluefish invade the back bays on all running tides. The best spots: Margate Bridge behind Margate, the Longport Bridge, and Absecon Bay. A run of blackfish ("tog") also occurs in May, especially around the Sunken Barge in Lakes Bay and the Seaview Harbor Rock Pile behind Lockport. (The tog return in October and are caught in numbers around the pilings of the bridges, especially the Ocean City Bridge.)

Fluke fishing continues to improve in May and June, at Lakes Bay, Ship Channel, and all of the back bays and tributaries. The back bays also provide fine weakfish action in mid-summer, with Lake Bay Creek a particular hot spot. Super crabbing heats up with the weather, especially at Scull's Bay behind Margate.

The action on kingfish begins in late August and early September, especially at Risley's Channel behind Margate, the channel at Lakes Bay, Rainbow, and Ship Channel behind Ocean City.

Tip: Anchoring is a must for kingfishing. Go top and bottom or two hooks that drop below the sinker and chum with frozen ground bunker in a chum pot. The whole outgoing tide and slack low is best.

In late fall, the embarrassment of riches that is Atlantic City area fishing becomes even more exciting as stripers and bluefish charge into the rivers and bays. Egg Harbor River is an especially hot spot for stripers, especially when baiting with bloodworm on bottom rigs by day and artificials at night.

Needless to say, there are a number of areas for shore fishing, although most of what we have discussed here requires a boat. The public fishing pier on Longport Boulevard is a comfortable and often productive place to fish. Fishing is allowed alongside three of four bridges on the Ninth Street Causeway in Ocean City (although not from the tops of the bridges) and from the perimeter of the Longport/Ocean City Bridge.

Boat rentals are available in Margate, Ocean City, Somers Point, and at the Harbor House condos. Boat ramps for launching your own rig are found in Somers Point, in Ocean City, and in Northfield at the Margate Bridge Causeway.

84 Mullica River

Directions: Take Garden State Parkway to exit 52; then follow Route 542 west to the Batsto area to fish the upper river. Take Garden State Parkway exit 50 for the Swan Bay WMA and the Port Republic area for saltwater river fishing.

The sweetwater upper reaches of the Mullica River are known to a limited number of anglers for superb pickerel fishing. Anglers working the creeks entering the river for a mile or so upstream and downstream from the Batsto ramp regularly land chain pickerel in excess of 2 feet in length. However, it is the lower, saltwater Mullica River that is famous for fishing, especially for kingfish, large sand porgies, weakfish, blues, fluke, and striped bass.

There is much to be said for "all season" fishing of the Mullica. In the dead of winter, Collins Cove on the west side of the Parkway Bridge often yields spectacular catches of white perch through the ice. Even in open water in winter, perch anglers in this area of the river have a ball. Our top Mullica River authority, Nuncie Bruno, uses a high-low crappie rig tied with #4 hooks, baited with either live grass shrimp or minnows. He fishes perch through the coldest months of winter, but in March turns his attentions to stripers, which are especially responsive to bloodworms fished at dead low and the whole incoming tide. Upriver into Log Bay, many anglers take stripers on floating and popping plugs.

Around Mother's Day in May is the unofficial kickoff of the weakfish season in the Mullica. Smaller fish are the first to arrive, and larger fish appear in June and well into July. In ML's personal experience, shedder crab is the bait of choice for weaks here, but he's taken more than a few on squid strip on a top-hook non-weighted bucktail.

Tip: *To conserve shedder crab bait, cut a small section of a discarded nylon stocking. Insert bait into the mesh and sew it up. Squeeze the little bag of bait, stick your hook through it, leaving the barb exposed. The bait will stay on your hook, the squeezing will cause a modest chum slick, and weaks will eat.*

In early summer, bluefish and fluke arrive and the fishing is particularly good with cut bait at Graveling Point and between the #1 and #2 lights. There is usually a profusion of delicious blueclaw crabs throughout the summer and in the waning days of August, kingfish and large sand porgies join the perch and snappers in the mouth of the river.

In October eels flood the lower river; consequently live eels are a fine bait for stripers. A bit upriver, however, bloodworms work as well if not better for bass. Good striper action continues right through the end of the year.

There are no boat-rental facilities on the Mullica; however, Chestnut Neck has two ramps, and there is another pay ramp upriver at Lower Bank.

Regarding shore access, we especially recommend Collins Cove, whether for ice fishing or for casting from the bank when the water is open. The best part is there is no bad time to fish this wonderful saltwater river.

85 Great Bay

Directions: Garden State Parkway to exit 58 onto 539 south to Route 9; go right on Route 9, then left on Great Bay Boulevard to one of several approaches.

Great Bay is the perfect name for so fine a fishing spot as this one, especially for anglers interested in stripers, bluefish, fluke, blackfish, weakfish, and kings.

Good access to the bay is found at the Chestnut Neck Marina at Port Republic (see "Mullica River") or in the town of Mystic Island at Great Bay Marina, situated at the end of Radio Road. Once on the water, you may want to make directly for Graveling Point, at the juncture of the Mullica and the bay. This is especially a springtime hot spot for stripers when the water temperatures reach into the mid-forties.

Figure 15 ■ Great Bay offers bigger blackfish than this one.
(Photo: Don Kamienski.)

Bluefish barrel into the bay in April, somewhat thin and ravenously hungry after a long swim north. The blues feed voraciously, primarily on small fish, right through mid-May before moving into the ocean for the summer. The secret to catching these blues rests in getting their attention, a feat at which well-known outdoor writer and lecturer Don Kamienski is a consummate master. Our tip here is courtesy of Mr. K.

Tip: *Try Don's method of saving herring caught each spring and grinding them up for chum. Pack a pound or so into individual plastic containers and freeze these homemade "chum logs" for later use. When bluefishing, place the chum log in a typical chum pot, secure it with a length of line, and drop it overboard—not resting but rather standing up on bottom so that it will lift and fall with the movement of the boat. This steadily releases small but attractive particles of chum.*

Rig with long shank, size 3/0 hooks and use enough of a sinker to hold bottom. Bait with a chunk of herring no larger than 2 inches in diameter, matching your bait to the smell and taste of the chum. To entice blues further, cut up some extra herring into very small chunks and if the tide is not moving super fast, drop them right over your own baited hooks—

just as Don does. If there are any blues in the area, this system will be irresistible to them.

Fluke compete actively with the blues for this cut bait. The last time ML and Don fished here together, they boated six fine bluefish but also caught more than twenty fluke on less than ideal fluke offerings—especially since fluke generally like moving bait.

If fluke are your primary target in the bay, look for a slow-moving tide. Grassy Channel is a first-rate area for fluke, as are the flats off the Mystic Island clam stakes.

Blackfish are taken in May opposite the fish factory along the sod banks, especially on crab baits fished on a flood tide. The deep, rough bottom in this area may also produce your largest fluke of the day. In addition, the bay is a virtual cornucopia of blueclaw crabs for handliners and trappers alike.

Sea trout (generally called weakfish by Garden State anglers) also are avidly sought in the bay, second only to stripers. Shedder crab is the best bait and a modest moving tide the best condition for catching weakies.

Another popular quarry is the kingfish, found in the bay in numbers in August and September. To catch them, anchor in the sloughs behind the fish factory and use bloodworms for bait. Remember that kings have very small mouths, so go with little hooks: #4 down to #8.

For linesides and weakies, blackfish and bluefish, kingfish and crabs and doormat-sized fluke, it's a Great Bay indeed.

92 Cohansey River

Directions: Take Route 623 south in Cumberland County to its terminus at the river, just below the juncture with Route 607.

The Cohansey River (some style it Cohansey Creek) is a beautiful tidal stream flowing through the South Jersey marshlands. Like its sister river the Maurice, the Cohansey provides one delightful surprise after another as it flows from sweet to brackish to salt. Named for the Delaware Indian chief Cohanzik, who made the river basin his headquarters and personal hunting grounds, the river is wreathed in legend. Black Beard and Captain Kidd are said to have buried treasures in the upper meadows of the Cohansey during the era when they terrorized shipping along the New Jersey coast.

At its upper reaches in Salem County, it is a fine trout stream, with more than 1,000 brook and rainbow trout stocked above Bridgeton from the dam at Seeley's Lake downward around Sunset Lake. Of special interest to us are huge carp that often can be seen in the vicinity of the Route 49 bridge. (Such large carp hooked in moving water promise an excellent fight.)

Tip: *Carp are among the wariest of fish. It is almost a rule of thumb that if you can see them, you won't be able to entice them to take your bait. Approach visible carp as carefully as you would the most wily trout in a high mountain trout stream and present your bait as carefully as you would a dry fly.*

Migrations of spawning fish into this tidal water provide additional diversity. Adult shad and herring run upriver in spring, followed closely by feeding striped bass. The herring are regularly caught on small gold hooks and are worth the effort, since they are superior baits for big stripers. (Smaller bass, caught year-round in the Cohansey, show a marked preference for bloodworms and grass shrimp.)

Snapper bluefish, weakfish, and occasional fluke also migrate upstream from the bay, and all are attracted to similar live baits, especially bloodworms, grass shrimp, and minnows. (Many anglers net their own minnows and grass shrimp at the mouths of feeder streams along the tidal Cohansey.)

The big channel cats patrolling the tidal water are partial to shedder crabs, although virtually any "stinkbait" will get their attention.

Ultimately the Cohansey drains a 100-square-mile basin in Salem and Cumberland Counties. Where it finally flows into Delaware Bay, the fishing for spot, croakers, bluefish, fluke, and big striped bass give the river its final twist of variety and excitement. All in all, this 31-mile river is not to be missed.

93 Fortesque

Directions: From Route 49 at Millville, turn onto Route 533 into Fortesque.

Fortesque styles itself the "Weakfish Capital of the World," a sobriquet which described the feverish fishing for weakfish in the late 1970s and early 1980s, when parking lots and boats were always full and anglers returned to shore with coolers full of weaks. Indeed, the state-record weakfish, an 18-pound 9-ounce monster, was caught in the bay off Fortesque.

Fortesque is located almost in the geometric center of the entry points of the Cohansey and Maurice Rivers into Delaware Bay. The weak-fishing remains a magnet to anglers today, from May right through October, although fishing for this particular fish has declined a bit more each year since the end of the 1970s. However, the fishing is fantastic for huge bluefish and stripers in November, especially inshore a mile off False Egg Island Point and in the Maurice River Cove, from East Point to Bug Light. The action is often fast and furious in the cooling waters of late fall, with blues and stripers taking cut bunker chunks or whole bunker heads.

The action is slow in the winter months, but in April, herring, perch, and undersized bass begin to appear, and by month's end, bluefish begin to be caught (especially from the Southwest Line to buoy 1) on mackerel and bunker chunks. The marinas at Fortesque come alive on May 1, when weakfish, fluke, and high-humped drum swell the action.

Tip: Fish for drum from an anchored boat at the top of the incoming tide, especially at the upper part of the Sixty Foot Slough and the "Pin Top." Bait with a big chunk of skimmer clam, sweetened with shedder crab.

As if to justify Fortesque's nickname, school weakfish averaging 1 to 3 pounds show up in July. Through the hottest weather of summer, the entire Maurice River Cove, from .5 mile to 1.5 miles out, is full of weakies. Generous numbers of flounder (fluke), spot, sand porgies, kingfish, and occasional throw-back stripers also invade in warmer weather.

In October, fishing for weaks, blues, fluke, and kingfish begins to slow. At this season, an enormous run of small yearling drum make their appearance off of Fortesque, not to be seen again until they return to spawn many years later.

Boat ramps are available up the Cohansey River, at the mouth of Fortesque Creek, and at Bivalve on the Maurice River. Shore fishing spots include Gandy's Beach, the bulkhead at the south end of Fortesque, and the end of the road outside of Heislerville at East Point Lighthouse.

Historical note: "The Weakfish Capital of the World" was once known almost exclusively for its croaker fishing. In the 1920s, the croaker fishing is said to have kept ten hotels operating in Fortesque to accommodate anglers from Philadelphia and New York. For this entire century, Fortesque has been known as a recreational fishing capital of rare opportunity, and with good reason.

94 Maurice River

Directions: Take Route 49 into downtown Millville. Excellent fishing access at Route 49 bridge. Take Route 49 west, then Sharp Street north to reach the Union Lake Dam.

The Maurice (pronounced "Morris") River changes along its 15-mile course from fresh water to brackish to salt as it travels from the foot of Union Lake Bridge into Delaware Bay. With a tide change of 5 to 6 feet, the river offers an embarrassment of riches to the angler, who needs a freshwater license in the upper reaches below the dam and at Port Creek but no license in the final miles as the river turns to salt.

Three fish dominate the action in the Maurice: herring, striped bass, and white perch, although yellow perch and sunfish also are caught in spring around the creek mouths, and big channel catfish vie with the stripers for cut bait below. Each year, large schools of mature alewife and blueback herring travel upstream from the bay to spawn, usually in late March or early April as the water begins to warm. Mature herring of up to 16 inches are easily caught on small, shiny lures, jigs, and even bare gold hooks.

Tip: Try small gold spoons on three-hook rigs with a quarter- to half-ounce *Dipsey sinker. Often you will hook two or three fish at once on this rig.*

Herring pile below the Union Lake Bridge where they attempt (unsuccessfully) to negotiate the fish ladder. Fishing action is hot here and the area is easily accessible, with plenty of good parking.

The herring are not great table fish, although the roe is quite palatable. However, they are valuable as baitfish for stripers and as bait (either whole for tuna and bluefish or in strips for fluke). For us, the principal benefit of the herring runs has more to do with the big striped bass that inevitably follow. Schools of big stripers tend to congregate in deep holes in the vicinity of Millville—some in the 15- to 20-pound class. The best time to take them is just at the end of the herring spawning run. Baiting with live herring on a big float is a natural and successful approach; however, big stripers also are taken on sand eels, bloodworms, and a variety of artificials, including jigs.

The best of the striper fishing takes place in May, all the way up to the dam. Cut fresh herring and live eels produce fine results here at the Silverton Boat Factory area. Stripers remain active through June, with cut bait in the river's upper sections producing fine fish. Significantly, big channel catfish take the same bait and produce additional fishing excitement in spring and early summer. The minimum size limit for stripers is 28 inches and the current creel limit is two per day.

Tip: When releasing fish, wet your hands and grip the fish in its lower jaw, *with the other hand supporting its weight underneath the belly. Do not reach into the gills, since stripers have razor-sharp gill plates. (It is never a good idea to reach into the gills of any fish to be released since fish are easily damaged by such handling.)*

White perch are avidly sought from Bivalve to the Maurice River Bridge, both from boats and from shore. The Port Norris Municipal Bridge, the pier at Mauricetown Park, and the site of the old Maurice River drawbridge are special hot spots for shore fishing, especially in early morning.

There is a modest shad run, generally in mid-April. Best action is experienced just at low tide near the Route 49 bridge. Green herring darts with green twisters produce 2- to 5-pound shad.

There are a number of "pay-for-launch sites" along the river, but the state also operates a good free boat ramp at Fowser Avenue in Millville. Convenient access for launching car-toppers can be found at the Mauricetown Bridge and at Dividing Creek. Boat fishing is our personal preference, but shoreline fishing is popular and very productive on the Maurice, especially in the vicinity of the Route 49 bridge.

At the river's terminus, anglers take spot, weakfish, blues, and summer flounder, and East Point Lighthouse, at the mouth of the river, is the site of great beach fishing. A 55-pound striper was taken at this point several years ago. It is fascinating to consider that in a 25-mile stretch, this river changes from a trout-stocked sweetwater stream to a lake to a great brackish-water fishery to a beach-casting paradise.

96 Sea Isle City

Directions: Take the Garden State Parkway to exit 17 at Ocean View. Follow signs to Sea Isle City; proceed south on local streets for Townsend's Inlet.

Sea Isle City is the area stretching from Corson Inlet to the north down to Townsend's Inlet to the south, and including Ludlum Bay.

When anglers discuss this rich and exceptionally accessible fishery, the word "fabulous" often crops up, especially when describing kingfishing in late summer. All of the back channels teem with kings at low water and some boats return to the dock with forty or more.

> Tip: *Rig for kingfish with two hooks that fall below the sinker but are held slightly off bottom with a small float. Anchor and chum with chopped bunker.*

Crabbing is also super throughout the summer, especially at the west side of Ludlum Bay on high tide. Crabbers who know how to feel for blueclaws use handlines with hunks of bunker, working back into the small channels where water moves slowly. Traps also work, but are less fun and less productive than handlines in the fingers of "pros."

A few early-season fluke are taken in the inlets in April, generally two hours before or after high and low tides. However, fluke and undersized bass come pouring into the bay and inlets in May, with the bass showing particular interest in bloodworms. Weakfish arrive in the back bays several weeks later, feeding almost exclusively in very early morning hours. Weaks are regularly taken on white bucktails with purple worms or clean lead-head jigs with purple or bubblegum-colored worms.

Fine fluke action continues in summer, especially in the deeper channels at Townsends Inlet. Small bluefish hit either bucktails or cut strip baits on top and bottom rigs.

In October, as finger mullet come into the back bays, the striped bass and bluefish action heats up significantly and continues well into the cooling trends of late November and early December. Live eels or black and silver Bombers fished under the bridges or around any other structure just after dark produces plenty of big linesides. Big blues charging into the channels behind the inlets are especially susceptible to Ava 27-type jigs without teasers.

Fishing access is excellent, including jetties and beaches to fish at the mouths of both Corson Inlet and Townsends Inlet. Fishing also is permitted from Corson Inlet Bridge. There is a public boat ramp across from Red Dog Bait and Tackle in Sea Isle City, and boat rentals are available in two sites in Sea Isle City and another near Townsends Inlet.

97 Delaware Bay

Directions: Garden State Parkway to end.

In our section on the Delaware River, we said there were four segments but we described only three. The fourth is Delaware Bay, one of the very best of our "100 Best" because of everything it has to offer. The bay is home at various seasons to an incredible array of fish species—often extremely large fish. It is all brine in content, even though the river brings plenty of fresh water down to the bay.

Shad and herring enter "The Big D" through this wide runway; then their fry make the long and treacherous trip past thousands of hungry mouths as they go back out to sea. However, striped bass, weakfish, blues, fluke, and drum are the main attractions in the bay, with plenty of porgies, blackfish, spot, and kingfish found in the part of the bay closer to Fortesque.

Until the 1970s, weakfish were the number one target of most anglers in the bay. Big weakfish as heavy as 15 pounds or more came into the bay each spring to spawn. However, an untold number of commercial fishing boats began netting the big sea trout before they could spawn, with predictable declining populations as a result. Now, however, government regulations have stopped this slaughter and the weakfish have returned. (Note: regulations favoring weakfish affect us garden variety anglers as well as the commercial netters, so be aware of the regs before you put fish in the fishbox.)

Sharks patrol the bay and often strike baits intended for other species. Anglers using spinning tackle have a brief but unfulfilling experience when this happens, since sharks overmatch light gear.

Among the largest and most powerful fish in the bay are drum. When anchored in the bay in a likely area, you can actually hear them drumming.

> Tip: *Use heavy tackle, at least 50-pound test line and a whole sea clam for bait. If a drum strikes, be prepared for a titanic struggle with a fish that may hit 75 pounds or even more.*

In the spring, stripers, blues, and weakfish feed best on outgoing tides, when the warmer upriver water is moving out into the bay. At this time, all three prefer bait to lures.

In the fall, big stripers—some as large as 50 pounds—feed actively in the bay, as do big bluefish. A whole head of a mossbunker is the blue-plate special for big stripers, with live eels a close second in preference.

Bay anglers enjoy excellent fluke fishing, often taking doormat-sized fish. Top and bottom hooks holding single strips of squid as bait frequently result in two fish at a clip, often with a weakfish rather than a fluke on the top hook. This is an instance of taking the good with the good!

Regarding access to the bottom of the bay, a 750-foot railway out into the bay is to be found in Bayview. This facilitates launching boats of substantial size, which are altogether recommended, since strong winds, thick fog, and heavy freighters heading in and out of the bay often make small boating a hazardous enterprise.

On Memorial Day weekend, an incredible flotilla of boats appears on the bay, some driven by "skippers" who are, to put it kindly, inexperienced, and to put it bluntly, downright rude. ML was fighting a big weakfish on light one-hand spinning gear in a Memorial Day crowd one year when a passing private boat cut his line as it sped by—no more than 10 feet away.

For Memorial Day fishing, we recommend you check out some of the other 99 spots described in this book.

98 Cape May

Directions: Drive to the end of the Garden State Parkway, follow signs to Cape May Point State Park.

The Cape May region is a virtual storybook of American maritime heritage, from Revolutionary War skirmishes fought with the British in its inlets and harbors to a gun battery built by the army during World War II to guard against attack. Cape May was a thriving whaling center during the seventeenth century and Lenni Lenape Indians used the area as their summer fishing grounds.

Today it remains a town of graceful Victorian architecture and quiet hospitality. Yet just off shore, an exciting, often spectacular fishery awaits the saltwater angler with a yen for excitement and perhaps a place in the record books.

The state-record barracuda, at 27 pounds 8 ounces, was taken off Cape May, as was the state-record greater amberjack, which weighed 85 pounds.

A state-record cobia, at 83 pounds, was taken here, as was the state-record (19 pounds 12 ounces) fluke. The record (9 pounds 12 ounces) Spanish mackerel was taken off Cape May, as was the record (880 pounds) tiger shark. Moreover, the long-standing 24-pound 4-ounce bluefish record was smashed here in late 1997, and the blackfish (tautog) record of 21 pounds 8 ounces—set here in 1987—was broken here in early 1998. And these records only begin to suggest the quality of the Cape May fishery.

Indeed, the entire coastline, from Avalon to Cape May, is rich in fishing opportunities, with Cape May Inlet providing a cornucopia of bluefish, weaks, kingfish, fluke (flounder), and stripers, especially when northeast winds pile up water, bait, and game fish along the shore.

Offshore, a massive run of mackerel generally starts around the end of March, continuing through April and into early May. The entire shore-line comes alive in early spring when weakfish, drum, fluke, and sea bass show up in increasing numbers and size. Less well known but no less exciting, big blues hit the sounds behind Avalon, Stone Harbor, and the Wildwoods in early May, with clouds of wheeling gulls announcing their arrival as they slash into schools of baitfish. The best action is in the shallow flats behind the barrier islands west of the main channels. Before the month of May is out, big kingfish also power their way into the shallows around Stone Harbor Municipal Marina and at the rocks in Hereford Inlet.

As the weather warms, the fluke action slows a bit and the weakfish move into the deeper holes in the back bays. But through the summer, sea bass, sand porgies, spot, blowfish, and small blues make the back bay a perfect spot to take the kids. In September, serious action resumes as small striped bass, spotted weakfish, bigger blues, kingfish, and huge fluke return to the bays and inlets. Late September finds the North Wildwood bulkhead producing striped bass, weakies, kingfish, and large numbers of blackfish.

One especially memorable September day, ML was murdering fluke at the inside end of Hereford Inlet on standard killy and squid rigs—and the fluke almost returned the favor! Fighting one 3-pound fish on the right side of the boat, a second rod suddenly jerked upward and over the gunwale as a second fine fish struck. ML grabbed this rod and had both fish on, just as the boat pitched and rolled, tossing him partway into the drink. Fishing buddy Lou Rodia raced to the stern, grabbed his shirt, and dragged him back into the boat—still holding both rods. And of course he landed both flatties!

From late spring to the arrival of cold offshore weather, shore fishing from local piers and bridges often promises surprising action. The best include a small municipal fishing pier at the west end of 83rd Street at Stone Harbor, the neighboring 96th Street Bridge, the Grassy Sound Crabbing and Fishing Pier just east of North Wildwood, Dad's Place at Ocean Drive, Grassy Sound, and the bulkhead at North Wildwood right at Hereford Inlet.

> Tip: *Crabbing and clamming also are big along the inland waterways touching these stretches of salt. Equipment for crabbing can be as simple as a piece of string weighted with a bit of lead and bait. Drop weighted baits overboard and gently lift from time to time. Crabs feeding on the bait will hang on at least long enough to be netted at the surface.*

More sophisticated crabbers use versions of the commercial box trap or trot lines which require permits (available from licensing agencies of Fish and Game).

Boats for fishing and crabbing are available at dozens of marinas and liveries up and down the coast. Some of our favorites are found at Stone Harbor, at Ocean Drive, at Grassy Sound north of North Wildwood Boulevard, at Grassy Sound Channel, and at Pier 47 Marina west of Wildwood.

Surf and Rocks

All forms of fishing can be fascinating; however, none seem to inspire quite the fervor that surf casting does. Devotees of fishing the "suds" can be seen along the state's beautiful sandy beaches from Perth Amboy to Cape May, and around the cape to the Delaware Bay, casting far out into the surf in solitary quests for large fish. Serenaded by the cries of gulls and pounding waves, these hardy souls fish some of the richest stretches of shoreline—especially for striped bass—to be found anywhere along the Atlantic Coast.

32 North Beach

Directions: Garden State Parkway exit 117; then take to Route 36 east, following signs to Sandy Hook.

North Beach is not for the angler who is unwilling to walk a bit. However, it is a superb location for fluke, striped bass, and bluefish, and well worth the effort. Park your car as far north as possible in the park and walk due northeast to the water. On your walk back through the sand, you will often be encumbered by the weight of a mess of fish.

Storms and tides change the angle of land severely in this area, but in normal weather, fluke are caught in large numbers from late spring through the end of summer. While surf casters typically prefer very long rods for long casts, no such gear is needed in the surf at North Beach. A steep drop to 20 feet or more of depth occurs very quickly here, a relatively short cast from shore.

"Flutter belly" is a lethal fluke bait here, and it was here that ML was taught about its use by the late Joe Sobel, a well-loved former teacher at Somerville High School. In filleting fluke, Joe would save and preserve as bait the unusable belly portion of the fish, a strip a half to one inch wide of uneven (fluttery) skin and meat located at the joint with the fluke's fins. ML has used this bait here and in other ocean spots with great success. (In some instances, this bait is prohibited, so be sure of local regulations.)

Striped bass fishing is best here during bad weather. Anglers score excellent catches in colder water in the start of easterly "blows," because as the waves kick up sand, stripers like to feed. Plugs that look like herring are deadly here, but some experts swear by spoons. (Bring your foul-weather gear and waders to North Beach, because the fishing often heats up as the weather turns ugly.)

Super catches of bluefish are regularly reported by anglers casting jigs—the shinier the better. But bait, especially whole small mullet, cut bunker, and herring, can induce blues and stripers to hit.

> Tip: Carry a sand spike for your second line when baitfishing and never forget to leave the drag wide open. Otherwise a big bass or blue or even a "surprise fish" is capable of taking your outfit out to sea.

Once, for example, ML was fishing a squid-strip and live-killy combination on a second bait-casting outfit. Suddenly the outfit was wrenched toward the water. Grabbing the rod just inches from the surf, he struck and watched line melt off of a screeching reel. The fish turned out to be a 10-pound blackfish, the largest any of the locals remembered being taken in the surf at North Beach.

North Beach is just up the coast from another local attraction, which ML learned about on a private boat while fluke fishing with four Somerset County police officials. Concentrating on the drift, ML glanced up from time to time to note an acute interest among his fellow anglers in sighting what he assumed to be land bearings, through good binoculars. Why else, he wondered, would they spend so much time peering shoreward. It turned out that the landmark to the south was Nude Beach, populated that day with plenty of sun worshippers. (Thinking this amusing, ML made the mistake of writing about the "land bearings" in his fishing column for a local newspaper. It turned out that the spouses of Somerset County's "finest" were somewhat less amused, a fact made clear to him by his erstwhile fishing partners thereafter for weeks on end.)

North Beach (not Nude Beach) is a surf-casting paradise, for short casters and long.

51 Marine Place

Directions: Garden State Parkway to exit 105 to Route 36 to Long Branch. Go 1 mile south on Ocean Boulevard then two blocks past the Roseld Avenue light.

Marine Place offers first-rate fishing in three different venues. First is the smaller rock jetty at the Deal Esplanade. Second is the longer jetty at Marine Place. Third is the beach/sand sections at or around both jetties, which are especially attractive to surf casters. All three offer fine possibilities.

We recommend parking at the Deal Esplanade. The small jetty there merits careful consideration, especially on any moving tide. A short walk to the south brings you to the considerably larger jetty at Marine Place. This jetty once was L-shaped, but the end was carried away in a storm, leaving a long, straight rock structure much beloved by anglers.

Joe La Presti of Stevens Bait and Tackle, an authority on fishing the surf and rocks here, suggests live bait or chunks—bunker or mackerel—for bass in spring, since stripers tend to move slower in colder waters and are not inclined to chase a fast-moving artificial. However, in summer and fall, lures that look like herring are just the medicine for big bass and blues. (Even in late summer, live eels still may produce the largest stripers.)

When fishing the jetties, two precautions are important. First, be sure your footwear is adapted to keeping footing on slippery rocks. Second, take a fishing companion, even if you generally prefer solitary angling. Even with corked or cleated soles, a nasty fall can happen, and without a partner to help you, a broken leg out on the rocks can spell disaster.

Casters working the sand often catch big fluke in summer, and sometimes the best action on bass or blues is along the sand—away from the rocks—as well.

A note on access and manners: Access is convenient from city streets, but a maximum of two-hour parking is permitted in summer between 8 a.m. and 6 p.m. The city fathers keep the beaches and jetties accessible, and in return it is important that anglers take all litter and debris with them, so we don't wear out our welcome.

64 Manasquan Inlet

Directions: Garden State Parkway to exit 98 to Route 34 south; then Route 35 south over the river to the Broadway jug handle to the inlet.

Where the broad Manasquan River meets the sea is an unparalleled fishing spot for "jetty jockeys" who flock from cities throughout the state to fish from the inlet's famous rocks.

Most people gravitate to the south side of the inlet at Point Pleasant and walk to the easternmost part of the rocks to get their lines in the water. The southern seawall was lengthened in the 1980s with odd torpedo-shaped rocks that can be especially slippery. The Point Pleasant side has a rare handicapped-accessible angling spot. ML met a wheelchair-bound angler happily fishing from this spot several years ago and has wondered ever since why more such sites aren't provided.

Slightly less popular but still well attended is the north end of the inlet on the Manasquan side, where the stout of heart (and steady of feet) walk out to the end of the often-slippery rocks of the jetty. Boaters heading out to sea each morning see many anglers on the rocks, most of them arrayed in good waders equipped with metal creepers on the bottom for footing. (Fishing from the north end's seawall is a lot easier and safer, although not necessarily as productive.)

Most "surf and rock" anglers prefer early morning, for the fishing as well as the visibility. Sunset also produces well and many of the real "pros" fish the rocks clear through the night, often with fine catches.

Tip: Carry a long "measured" stringer out on the jetty, marked with length measurements so you can check fish to make sure they are legal sized. Tie your stringer's dry end around a high dry rock so that it won't disappear under the incoming tide.

What kind of fish? False albacore are sometimes caught from the rocks, as are striped bass and bluefish. Make sure to have a Bomber plug or two in your tackle box for stripers and Ava-style jigs for blues. Weakfish also are taken on metal, especially in the quiet hours when boats are not busting in and out of the inlet, and blackfish are known to poke around the rocks hunting for small crabs. Anglers rigged for these "togs" often catch large ones. Plenty of blues and bass are taken in the inlet on live herring, bunker, or eels as well as artificials.

At times you will encounter divers at the easternmost sections of the inlet, spear fishing for stripers and blackfish, looking for lobsters, or just poking round. Of course, carefully avoid them and they should return the favor.

68 Normandy Beach

Directions: Garden State Parkway to Route 35 south, pass Thunderbird and park in Brick Beach Lots 1, 2, or 3 (check for permit requirements).

Normandy Beach is a name that lives large in the history of World War II's European Theater. Combatants also storm ashore in New Jersey's Normandy Beach, but the players here are striped bass, often of exceptional size and strength. In fact, the striper fishing is so pervasive here that it tends to obscure the other fine sport fish that abound in the surf.

Moreover, if a version of *The Old Man and the Sea* were written about striper fishing in New Jersey, we are certain the setting would be Normandy Beach and the protagonist would be Ernie Wuesthoff, who may have caught more striped bass than any other man alive. Ernie has been in the bait-and-tackle business (his "Bait 'n Tackle" Shop is located in the center island of Route 35) for more than sixty years but more often than not, he is found on the nearby beach rather than behind the counter. Ernie fishes only for stripers, scorning the many weakfish and blues that work the surf at Normandy Bay. As to conditions, his formula is simple: if he can walk and the wind and weather are not impossible to fish, he is "in the suds."

Ernie, who is often featured in Capt. Al Ristori's *Star Ledger* fishing column, is a study in how to surf fish. He favors a smaller rod and prefers to fish close in. In order to do this, he "reads the beach," looking for cuts, sand bars, washes, and other irregular bits of bottom carved in the sand. He knows this is where bait will be swimming, often with stripers in full pursuit.

The action at Normandy Beach starts about mid-April each year and sometimes continues through mid-January if the weather is mild and water temperatures remain moderate. The stripers are particularly voracious when sand eels invade the surf.

Tip: *If your lure comes back with a sand eel snagged on a treble, get seriously ready. If enough sand eels are in the surf that you snag one, the stripers are there too.*

For pros like Ernie, individual stripers may be cast to and caught. One day in 1997, Ernie Wuesthoff was chatting with a nearby angler who had made hundreds of fruitless casts, when he noticed and pointed out a bit of movement in the water. The other man scoffed at what was clearly just wave action, so Ernie made one cast—his first of the day—and dropped his yellow Bomber plug right into the striper's mouth. In the fight that ensued, the big lineside stripped line until there were only six turns left on the spool, necessitating Ernie's going right into the surf (in his street shoes) after him. He ultimately landed the fish, which weighed 28.5 pounds and is still referred to as the "one-cast bass."

Normandy Beach is one of the best surf-casting spots on the New Jersey coast, and we recommend it not only for the massive bass that cruise its waters, but also the possible opportunity to watch the consummate surf caster at work, if Ernie Wuesthoff is on the beach.

71 Seaside Heights

Directions: Garden State Parkway to Route 35 south to the boardwalk at Seaside Heights.

Seaside Heights offers anglers two entirely different fishing experiences without ever leaving land. The first is pier fishing; the second is surf casting. Both are often extremely productive for reasons having to do with structure and baitfish behavior.

Actually, two piers are visible at Seaside Heights: the so-called "Fun Pier" and the "Casino Pier." The Casino Pier allows anglers to walk out for a fee and fish straight into the ocean. Anglers catch fluke here in summer and also blues and stripers. When the weather turns cold, anglers often are rewarded with fine catches of ling at night.

> *Tip: When selecting a night for ling fishing, make sure that you will be fishing in a modest wind or a wind blowing out of the west. A westerly wind produces a calm and clean ocean right in at the beach and the best conditions for fishing from the pier.*

"Ling jerkers" fish with cut herring or a hunk of mackerel. Some years ago, the pier was a prime winter fishing spot for whiting, before the fish all but disappeared from these waters. When they reappear, as they are expected to do, the pier will be lined with nighttime anglers again, since the Casino Pier is one of very few such places in New Jersey for this type of fishing.

The main interest at Seaside, however, is not pier fishing, but rather surf casting for bluefish and stripers from the beach. Fishing at night between the piers is most productive, when blues and bass, searching for bait that hide around the pilings, occasionally venture out to feed. Writer/guide Joe Kasper finds superb action along the mile-long stretch

Figure 16 ■ Joe Kasper with a huge striper he caught.
(Photo: Captain Joe Kasper.)

from the northern end of Seaside Heights down to where telephone poles have been buried in the sand as a barricade. Riding to the action on his beach buggy, Joe catches large numbers of bluefish here on small metal lures and on plugs that simulate herring. When the water is boiling with blues on a feeding frenzy, he often fishes a popper on top for additional excitement on the strike.

A beach-buggy permit may be purchased at the municipal building in Seaside Heights, currently for an annual fee of $40. The beach buggy is handy for scoping out bottom formations at low tide as well as fishing them when the tide is in. Every storm brings new bottom formations, and while one big easterly blow can create new cuts and bars where bait and fish will tend to congregate, the next one can flatten them all out again. By hitting the beach at dead low tide, you can pick out high and low sand where the fishing will probably be best on the next high tide.

The best fishing generally occurs at the beginning of autumn and into the start of winter. There is a smaller spring run of bass and blues, a time when both are more susceptible to bait than lures.

74 Island Beach State Park

Directions: Take Garden State Parkway to Route 37 east to 35 south toward Seaside Park; go south to park entrance.

Island Beach State Park extends from Seaside Park to Barnegat Inlet and every foot of its beachfront offers fine fishing opportunities. And one of the all-time most successful anglers along this stretch is well-known writer and guide Joe Kasper, whose log book for 1997 provides an idea of what awaits you along this gorgeous beach.

In 1997 alone, Kasper landed in excess of 250 striped bass at Island Beach State Park, weighing up to a high of 30 pounds, plus another 250 bluefish reaching a high of 18 pounds. And that's not all. Friend Joe also landed 35 fluke and another 60 weakfish up to 6 pounds each, mostly caught at night. Add up these numbers and it is easy to see why Island Beach State Park is Joe's favorite surf spot.

Ready to try it? Here are some of the particulars. The park is open 24 hours a day, 365 days a year. You pass through a guarded gate on entering, and there are three primary exits onto the beach from the park. Exit 1 is called Gilikin's, and it leads to a section of ever-changing sandbars and ridges, and superb fishing. The second exit is labeled 7A and also leads to super fishing. Closest to Barnegat Inlet to the south is marked exit 27. You can park here and walk to the beach or take a four-wheeler straight down to the sand or walk to the North Jetty at Barnegat Bay.

Late each summer and through the fall, small mullet abound in the surf and trigger weeks of even better fishing than has already been enjoyed through spring and summer. Stripers and great schools of migrating bluefish storm into the surf in pursuit of the baitfish and the feeding orgy is on. Live mullet can be bought (while supplies last) in bait shops, or you can catch your own.

However, many of the real pros along Island Beach State Park stick to artificials. For stripers and weakfish, especially in the fall, the lure of choice is a 5- to 7-inch yellow and black Bomber. This lure is made even more deadly by tying a small nonweighted bucktail teaser 2 feet above the plug, on a short, stiff leader.

Tip: *Try this rig yourself. The plug at the end of your line appears to be a small fish chasing an even smaller fish (the teaser). ML once hit a "double header" of bass this way, with the lure chasing the teaser being an Ava 27 jig.*

Blues hit metal here, preferring smaller to larger lures. Instead of the typical Ava-style 47 lures commonly used on party boats for blues, go with the smaller 007 size (with no tube or twister on the end) or a small Hopkins spoon. Joe Kasper prefers a plain silver spoon with treble hooks removed and replaced with a single hook to avoid snagging.

The best time to fish the surf at Island Beach State Park is the top of high tide, until the first two hours of outgoing. Nowhere in New Jersey are beach buggies more popular than the Park. The state issues yearly four-wheel drive permits, at a fee of $125 each, to those anglers who follow the schools of fish up and down the miles of beach.

Figure 17 ■ A fine bluefish caught in the surf at Island Beach State Park.
(Photo: Captain Joe Kasper.)

Excitement? When the gulls start to wheel above the bluefish smashing into schools of bait, a short cast from where you stand, Long Beach State Park may be the most exciting place on planet Earth!

77 Barnegat Inlet

Directions: Garden State Parkway to exit 63 (Long Beach Island) and Route 72 east to the island. Go north to the end of Long Beach Boulevard. (Watch speed limit signs.)

The north side of Barnegat Inlet has been discussed under the heading "Island Beach State Park," since Island Beach's southernmost section is at the north side of the inlet. Consequently it is the south side of the inlet we are covering here, setting it apart as one of "New Jersey's 100 Best" because of its extraordinary striped bass fishing.

The landmarks of the south side are hard to miss, since they include a Coast Guard Station and a very imposing lighthouse standing guard for ships heading in from the sea. And the fishing—especially from April to December—is often fabulous.

Striped bass own the hearts and minds of the more serious anglers here, although fine catches of weakfish, blues, fluke, and blackfish, plus respectable catches of kingfish and winter flounder, are often rung up. However, more angler hours are spent fishing for stripers in Barnegat Inlet than for all other fish combined.

One of the best and most productive spots for fishing the inlet is the South Jetty, which is approximately three quarters of a mile long. The jetty is flat, comfortable, and easy to fish. A handrail runs along its first 1,000 feet (which helps prevent the awkward "green rock flop" that ML has demonstrated—to the considerable amusement of other anglers—on jetties and in streams up and down the entire east coast). The jetty was completely rebuilt in 1992 to correct conditions so dangerous that in the 1980s, some insurance carriers would not provide insurance for boats sailing out of the inlet. By extending the jetty in the rebuilding process, combined with a massive dredging effort, both the safety and the fishing in the inlet were significantly improved.

Barnegat Inlet is one of the striper hot spots selected above almost all others by author/fisherman Frank Daignault, who gave it a four out of a possible five rating. He suggests, and we concur, that bass bite best at the far end of the jetty on low slack water.

> *Tip: Also fish for kingfish on the last hour out, slack, and the first hour of incoming tide. Use supermarket shrimp for bait. The best fishing will probably be in September on weekdays, when boat traffic is light.*

When the tide is not "cooking" too fast, fish inside the inlet for bass. But if it is really cranking, try outside, into the ocean itself, and around the rocks, which harbor many fine blackfish. Weakfish will generally be found on the outside as well, generally very early or very late in the day. They are particularly susceptible to high-low rigs baited with skinny, tapered squid strips.

Ample parking is available, and with your vehicle close at hand, you may want to drive back to Eighteenth Street and check out the day's catches made by the large fleet of party and charter boats returning to the docks.

80 Loveladies/Coast Avenue

Directions: Garden State Parkway to Long Beach Island (LBI). Go north on Boulevard to Loveladies and right onto Coast Avenue.

The largest stripers weighed in at Fisherman's Headquarters each year generally exceed 50 pounds. According to manager Brent, these fish are often taken in the surf off Coast Avenue.

Productive surf fishing begins here in April. Live herring is the bait of choice, but bunker chunks work well, too. (Perhaps stripers on chunks sounds odd; however, RB took the first striper he ever saw on a chunk of mullet fished dead on the bottom—for catfish—in the massive Santee Cooper impoundment in South Carolina.) The best fishing is generally on the last two hours of an incoming tide and the first two hours outgoing. Depending on the presence of forage fish in the water, stripers are hooked as close in as five yards from your feet and as far out as you can cast. Thus we recommend taking two rods and sand spikes to this site: one for fishing close in and the other for "slinging lead."

> *Tip: Make sure you use a "fish-finder rig"; that is, a rig which holds your 2- to 4-ounce pyramid sinker on bottom but allows the long-snelled hook to slide freely up and down the line to avoid spooking your fish on a bite.*

As spring turns to summer, bluefish, fluke, and sunbathers move into LBI. The presence of fluke is undependable, but if the surf is warm and no south wind is blowing, chances are good that a pile of "flatties" are close at hand. Brent tells us that rigging a frozen mullet on a fluke rig and retrieving it slowly through the surf seems to produce most fish. Add a strip of squid for extra action. Bucktails also produce good action for blues as well as fluke, and occasional weakfish.

LBI becomes incredibly crowded on hot summer days, so it is important for anglers to avoid typical sunbathers' hours. Parking is legal on Coast Avenue, very close to the action, but it is important to avoid blocking residents' driveways. Beach access also is available for four-wheel-drive surf buggies.

If LBI translates to fun in the sun for New Jersey sun worshippers, it spells great surf fishing, big fish, and convenience galore for the rest of us.

90 Atlantic City Surf and Rocks

Directions: Garden State Parkway to Atlantic City Parkway into Atlantic City to Oriental Avenue, which brings you to stretches of beach adjacent to the inlet and the T-jetty.

Every since Albert McReynolds smashed the New Jersey state record for striped bass more than ten years ago at the Vermont Avenue jetty in Atlantic City, thousands of anglers have traveled to this famous resort city to "gamble" that they might create their own records. McReynolds's massive bass weighed 78.5 pounds and may never be beaten on rod and reel in the Garden State.

Fishing at the Atlantic City area is excellent from the sand or from any of the jetties located along the shoreline or from the inlet jetty itself. Night is the best time to fish for stripers, especially in the fall. (Please note: in the interest of safety, we strongly caution against fishing alone at night.) Boat traffic is often heavy around the inlet. We advise you to wait until it slows before walking out on a jetty and slinging lead. Be sure to bring a heavy-duty break-proof flashlight along.

> *Tip: The best bait each fall is, in order, eels, eels, and eels. When casting from the rocks, keep your reel wide open or your drag very lightly set. The bite is sometimes a hard slam, but at other times, barely a "pick-up." Let the bass swim away with your bait for at least 5 feet, then slam the hook home!*

The T-Jetty is for less nimble-footed citizens, but can still produce nice action at times. It is weather worn and therefore easier to traverse.

There are plenty of bluefish mixed in with the bass at Atlantic City. However, they are best sought with metal like a Hopkins spoon. A bunker chunk is an alternate for both bass and blues, especially for anglers who are less inclined to cast (and cast, and cast).

Atlantic City is a unique spot where you can play by day and fish by night, or vice versa. Not at all a bad parlay, especially with record stripers out there in the surf.

The Mighty Atlantic

There are famous offshore fishing spots like the fabled Hudson Canyon, the Mud Hole, and Manasquan Ridge where giants swim the depths and wait to challenge anglers' skill, strength, and endurance. An 880-pound tiger shark, a 1,046-pound blue marlin, a 1,030-pound bluefin tuna, a 530-pound swordfish are among the record leviathans taken in these waters. Licensed boats and skilled captains and crews take anglers out to these special places in the sea, and for all of us, it is difficult to leave the dock without a heart-pounding sense of suspense and excitement at what lies beyond the horizon.

37 Seventeen Fathoms

Directions: Travel southeast of Sandy Hook from the Atlantic Highlands Municipal Dock, to an area 2 miles north by northwest of the B. A. Buoy.

Seventeen Fathoms is an ocean hot spot in the general vicinity of the B. A. Buoy and the Mud Hole. In fact, the eastern edge of "Seventeen" reaches into the deeper Mud Hole water.

The name Seventeen Fathoms derives from its depth of just over 100 feet. (A fathom is 6 feet. Dividing fathoms into depth here yields the name.) Of course, this depth recommends stiff-action rods for properly setting hooks. "Seventeen" is a vast area of rocks and snags, all very attractive to marine life. It acts like a magnet to fishing boats from Long Island, Sheepshead Bay, Atlantic Highlands, and Belmar. When the water begins to get cold each fall, the area comes alive with blackfish and schoolie cod. Ringers hedge their bets by rigging a blackfish hook or two at bottom,

Figure 18 ■ Anthony R. Monica with his state-record blackfish caught in 1998 on the *North Star* off Ocean City, N.J.
(Photo: Al Ivany, N.J. Division of Fish, Game, and Wildlife.)

Figure 19 ■ A real Jersey/Florida "pro" with a Seventeen Fathoms blackfish. *(Photo: Captain Joe Kasper.)*

baited with green crab bait, and a size 7/0 codfish hook 5 or 6 feet up the line on a 2-foot leader, baited with a whole skimmer clam, for cod.

Another surefire "tog" and codfish bait is conch. On Election Day, 1951, ML took the pool on a Sheepshead Bay headboat with a 20-pound cod, taken on tenderized conch meat. It works as well to this day.

> *Tip: Whatever your bait, it is important to check it from time to time, even if fishing well above the rocks for blues but more especially if fishing at bottom. The reason: huge schools of talented bait thieves called bergals hide in the snags at bottom, waiting to pick off your bait before a more desirable fish can get to it. (One reason blackfish experts prefer crab or conch baits is that these baits are more difficult to steal.)*

Bluefish are caught in large numbers at "Seventeen," especially at night. The blues taken here are generally bigger than those taken closer to shore. Big blues, big blacks, and schools of cod make Seventeen more than worth the ride.

38 Scotland Buoy

Directions: The Scotland Buoy is located approximately 3 miles offshore of the old Coast Guard Station and the famous Sea Gull's Nest Tavern on Sandy Hook.

In describing the B. A. Buoy area, another offshore hot spot, we note that a red ship named *Scotland* once served as a marker and navigational aid at the head of Sandy Hook Channel, and that it was replaced with a buoy of the same name. In fact, the Scotland Buoy marks more than shipping lanes; it marks a superb fishing area about one square mile in size and loaded with fish.

Longtime party-boat skipper George Bachert, who has run several party boats out of Atlantic Highlands Municipal Marina for more than twenty years, often takes his fares to this location for wonderful bottom fishing. In fact, more than 100 wrecks and other structure litter the bottom, attracting the crustaceans and small fish that bring larger fish to the area by the thousands.

Figure 20 ■ Ling like this one are commonly caught at the Scotland Buoy. *(Photo: Captain Joe Kasper.)*

Back when whiting reigned supreme, we caught these fish in huge numbers on the Scotland Grounds, during daylight hours and at night. Today the fish of choice are blackfish, porgies, sea bass, and ling. Many bluefish and striped bass are caught here as well. Indeed, many anglers pick up "beast-sized" stripers when trolling bunker spoons here in the fall.

Even though "Scotland" is so small an area, the depths vary from 40 to 80 feet.

Tip: If you are after ling in the springtime, try a square chunk of fresh mackerel as bait. Your ling count will go up with this super bait, especially if pesky bergals are not in the vicinity. Herring is good too, but cut in sloppy chunks rather than fancy strips.

41 B. A. Buoy

Directions: The B. A. Buoy is located 16 miles southeast by south of Atlantic Highlands.

The B. A. Buoy marks the beginning of a vast area famed for rich offshore fishing and deservedly, among the most prized of our best ocean spots. Captain Ron Santee, Sr., of Atlantic Highlands, an authority on the area, told us that the initials B. A. stand for Barnegat Approach. When his 65-foot party-fishing boat, *The Fisherman,* sails out of the Atlantic Highlands Municipal Marina, it often heads for the buoy, a 145-degree run in a southeast by south direction. The ride of 16 miles takes an hour and a half at normal cruise speed.

Along this trail, too, are the Scotland (to the west) and the Ambrose (to the east) Channel Buoys. From the buoy south, this area is an enormous shipping lane. In fact, ships bearing these names had been painted red and for many years were anchored as stationary markers for captains to guide their cargo ships. (See "Scotland Buoy.")

Fishing in this area is often nothing short of spectacular. In the area of the buoy, in depth averaging 120 feet, ling are a primary target. For much of the year, they are commonly found in numbers on both the soft, open bottom and around the many wrecked boats that lie in the mud.

Big blackfish and sea bass also are caught here in exceptional numbers. Occasionally, cod and pollock are found on the harder bottom sections, especially from mid-winter to the middle of spring. Interestingly, once a break-up of Hudson River ice occurs, the sea water is dramatically cooled by huge chunks of ice charging out to sea. Shallower water gets cold more quickly, of course, but even at the B. A., temperatures sometimes drop below the comfort zone of most of these fish. A mild winter can see this area producing action virtually nonstop, but a heavy ice melt often signals the end until April.

Back when whiting were more plentiful, the season began in earnest in November in the Scotland/Ambrose areas, but as the waters cooled, they moved to the deeper, warmer waters at B. A. Thus the "winter fishing" for these "frostfish" actually began in March or April around the B. A. Buoy.

> Tip: Since you are fishing in deep water, leave your light, whippy rods home. You can use fairly light line (20- to 30-pound test monofilament), but a 7- to 8-foot boat rod, or an extra-stiff bait-casting or spinning rod will allow you to set your hook effectively.

Crab bait is best for blackfish and a strip of herring often produces the best action with most other species, especially whiting and ling. Skimmer clam also works well. Listen to your boat captain for best results.

Captain Santee reminds us that tuna used to be caught in the area of the buoy on a regular basis, especially schoolies weighting 20 to 40 pounds, but occasional giants of 750 pounds or more as well. As recently as mid-December 1997, some anglers were finding that the ling they were reeling up were being gobbled by monster tuna. Something to think about when cruising out to fish the B. A. Buoy.

ML especially goes back a long way with this area. One day in the distant past, when stationed on a Navy destroyer escort returning to port after a cruise, he noticed that *Ambrose* was no longer in its accustomed anchorage. It turned out that the ship had been rammed by a larger cargo vessel in a heavy fog. Eventually, both the *Ambrose* and the *Scotland* lightships were replaced by safer marker buoys.

44 Mud Buoy

Directions: The Mud Buoy is located roughly 6 to 7 miles offshore, due east of Sea Bright.

The Mud Buoy is an area used for dumping clean dredge spoils and other materials and the name of the actual buoy that marks it. The location of the marker buoy is not stationary; it is moved each time the dumping site changes. Captain Ron Santee tells us that it is stationary now, at least for the time being, as dumping activities were suspended late in 1997. Of interest: this closing resulted from protests of the dumping to government. It may or may not remain closed in the future.

For the past several years, boats have come to the Mud Buoy from a dozen or more ports to get in on the superb bluefishing. In fact, many anglers have come to regard this as the best bluefish grounds of the 1990s.

Water depth here varies from 35 to 60 feet, with some of the shallower areas accounted for by significant dumping. The outer edges of the site are as deep as 80 feet. Interestingly, the dredged materials attract massive numbers of sand eels, rainfish, and squid—all attractive to blues.

Tip: For bluefish, especially the 1- to 3-pounders, use a single hook, without wire leader. Yes, you will be cut off occasionally by the "choppers," but when fishing in a crowd, your offering will look far more natural and your bait easier to take without wire.

Boats fishing this spot do better during the day than at night. Their primary targets are blues, but anglers also pick up nice catches of sea bass and blackfish along the area's rock-strewn bottom.

50 Shrewsbury Rocks

Directions: "The Rocks" begin only a short distance offshore of Monmouth Beach and extend to the east for 3 miles.

The Shrewsbury Rocks is a noted fishing spot for blackfish, sea bass, fluke, and bluefish. The big bonus, less well known than the other species, are big early-season striped bass.

True to its name, the bottom of this spot is strewn with large rocks. This creates a predictably attractive feeding ground for blackfish in the spring and fluke in the summer. Vast mussel beds and many snags also adorn the bottom and, like the rocks, have a tendency to cost anglers their terminal rigs.

Figure 21 ■ The Shrewsbury Rocks hold plenty of porgies in the fall.
(Photo: Captain Joe Kasper.)

Bluefish also flock to the Rocks in large numbers and mackerel stop to feed on their winter migrations. Seabass and porgies also are caught here and the marine life is swelled by swarms of bergals, which torment anglers fishing bait at or near bottom.

Tip: *Try extra-large baits for fluke. A whole squid is too large for sea bass and bergals, but just right for "doormats."*

The Rocks hold stripers very early in the season. While most anglers are fishing for blues or awaiting the bunker run, a few anglers score great catches of linesides here, usually jigging with bucktails or trolling with bunker spoons.

This spot is especially memorable for an event that took place more than 40 years ago involving a very small fish. ML and six companions were bluefishing from a 26-foot rented boat when the engine exploded. All seven anglers made it into the water and were rescued by other boats. On board a rescue craft, a bit of comic relief was provided when ML's late father took off his shirt. A little killy—the first catch of the day—jumped out and landed on deck. Somehow it had managed to swim into Harry Luftglass's shirt.

53 The Farms

Directions: The Farms are located 10 miles east by northeast of Asbury Park.

Boats from a number of ports, especially the Atlantic Highlands, Belmar, and Point Pleasant, make for The Farms, and often come back to their docks with super catches. This wonderful fishing spot marks the western edge of the Mud Hole and the beginning of a steep downward slope in depth. One of the significant features of the Farms is its hard rocky bottom, as distinguished from the sand found inshore or the soft, gooey bottom in the Mud Hole. The water is 90 to 100 feet deep; consequently you'll need a stiff rod to sink the hook into bottom-feeders.

Cod are often found on the Farms in the spring and the fishing is even better in the late fall. It was in the late fall of 1964 that ML found himself on board the *Captain Joe* out of Belmar—a day made memorable by his catching the only two codfish taken the entire day. Just luck?

Blackfish anglers flock to this area on significant temperature changes, when the bitterly cold water starts to warm each spring and in December, when shallower water becomes too cold.

Captain George Bachert feels that the Farms is an underfished spot for big sea bass as the weather starts to get colder. He tells us that big bass stop over to feed here as they make their trek eastward to deeper ranges out to sea. Mackerel and blues pass through the Farms in spring and fall, along with herring.

Huge numbers of bluefish congregate to feed every summer and when the action is hot, it's difficult to keep bait in the water.

Tip: If blues are not eating aggressively, you can tempt them by catching little bergals at bottom on small hooks. Hook a bergal, let it out in the chum slick, and brace for a strike. Generally you won't have to wait for very long.

57 Texas Tower

Directions: Sail 67.5 miles east by southeast out of Belmar Inlet.

Say the word "Texas" and exaggerated size comes to mind. In the case of the Texas Tower, this is wholly appropriate—not only because of its large debris field but also the size of many of the fish taken in the 180-foot depths at this hot spot.

The Texas Tower grounds are named for an oil rig of the same name that once loomed high over the horizon in this spot. In the mid-1950s, the tower broke up and fell to the ocean floor, creating a debris field much larger than a football field. Over time, this giant junkyard was augmented with additional debris, including two barges that were scuttled at the site, and all of this hard substrate material was colonized by mussels, mollusks, crabs, shrimp, small and then ever-larger fish. Superb fishing soon followed at this deepwater artificial reef.

Fishing at "the Tower" is excellent year-round, according to Captain Gary Fagan, whose Belmar-based *Big Mohawk III* is one of several headboats in the area rigged and equipped for deepwater fishing. (Coast Guard regulations prohibit most such boats from sailing beyond a 20-mile limit.)

Giant pollock and codfish are the featured fish in this location, especially in autumn and spring. Captain Gary generally sails early when he fishes the Tower, since the best fishing for pollock is usually in the dark before sunup.

> *Tip: At such times, a 10- to 17-ounce jig works better than a baited hook. Some intrepid anglers add a very large red or green tube teaser several feet above the jig, leading to a hysterical tussle when two huge fish strike simultaneously.*

Cod begin biting at dawn. Large white hake and their smaller cousins, red hake (ling), are taken here on bait, day or night. Incidentally, while white hake are unlovely to look at, they are fine table fare.

Anglers jigging for pollock between late August and the end of October are often stopped cold in their retrieves by bluefin tuna—generally "schoolies" in the 10- to 50-pound class. Regardless of size, tuna tear line off reels at incredible speed and anglers at the Tower do well to keep this in mind.

Two other species haunt the spot: pesky buck-toothed bergals that manifest incessant appetites for your bait, and sharks of every size and origin.

Is Texas big? No matter how you measure this "Texas," the answer is a ringing affirmative. Sharks, tuna, massive pollock and cod abound in deep water so far offshore that anglers completely escape boats, crowds, and the everyday cares of the real world.

59 Klondike Banks

Directions: Go 6 miles east by southeast out of Belmar Inlet.

The Klondike is a relatively small patch of real estate—about a square mile—that can be "pure gold" for anglers. Purportedly it got its name about the time of the Klondike gold rush up north. As the story goes, a bunch of pound-net fishermen rowed their heavy wooden boats well out to sea one day and were rewarded with tons of fish. Supposedly they likened their success to striking gold, hence the name. (We can't vouch for the accuracy of this old-timer's story, but it sounds good.)

The shallowest section of the Banks is only 48 feet, but it slopes down to 80 feet or more at its deepest edge. The shallow section is nearly all sand, but the deeper eastern section has lots of rocky bottom. The *Big Mohawk III* is one of the boats that regularly fishes the rocky section for blackfish, which run considerably larger than normal here, and sea bass. In the spring, the sea bass move inshore with ling mixed in, and both fish are taken in numbers. ML fished the Klondike aboard the *Capt. Bill Van* in the 1970s and 1980s, anchored up and catching blue after blue on wired, baited hook dropped into the chum slick. Even better were the fluke, which hit smelt at bottom every October.

However, the Klondike is actually best known for top fish. Blues are regularly taken here, and late in the summer, bonito and surging false albacore are caught as well.

Tip: If you get chopped off two or three times in a row while fishing for bonito, bluefish are attacking your line and you may want to change rigs, taking off the small tuna hook and spearing bait. However, even if every second hit is a "chopper," stay with the bonito, since the next strike may be a tasty 5- to 8-pound streak of dynamite.

Fluke swim offshore in the fall and stop to feed along bumps and ridges. When the first fluke is caught by anyone on your boat, drop a line down to the bottom rigged for fluke—without wire and, if possible, baited with a whole head-hooked smelt.

For all other fish, sand eels are the main forage species in the fall. If eels are spotted in numbers with fish-finding equipment, fasten your seat belt. It should only be a matter of time before big fish arrive to eat the eels, get distracted by the chum line, and come to your baited hooks.

Figure 22 ■ This nice false albacore was "mined" at the Klondike Banks. *(Photo: Captain Joe Kasper.)*

65 Point Pleasant

Directions: Garden State Parkway to exit 98 to Route 34 south; then Route 35 south over the river to the Broadway jug handle to parking at the inlet.

Looming out to sea from Point Pleasant are some of the largest fish in all of New Jersey's coastal waters. The state-record bluefin tuna, a 1,030-pound behemoth landed by Royal Parsons, was taken in these waters, as was the state-record 759-pound white shark, which was caught by Jim Kneipp. Less menacing, but no less a record, was the 8-pound 2-ounce sea bass taken off Point Pleasant on one of the many Bogan family boats.

All of the most highly sought coastal fish also are found off Point Pleasant, with special emphasis upon fluke, mackerel, bluefish, and striped bass. Rather than repeat a recitation of which fish during what seasons, we thought this a fit place to concentrate on one fish—fluke—and the best way to catch them, here and along the coast from the North Beach of Sandy Hook down to Cape May.

Beginning with tackle, you can get away with use of light spinning outfits, depending upon crowds (on party boats) and tides. We have often fished with medium-weight (one-hand) spinning outfits and lines of not more than 10-pound test.

We prefer light tackle because it allows us to feel the light bite of a fluke more effectively, and thus improves our ratio of fish to bites.

We recommend using as light a sinker as can hold bottom, with a long-leadered hook tied in above the sinker knot. The late Captain Charlie Selby of Wildwood taught us that the only hook to use is gold in color, and that a new hook should be tied every trip.

Tip: Fluke often swallow the hook, and thus it is important to keep a ready supply of extra unsnelled hooks at hand. When you catch an undersized fluke that has swallowed the hook, be careful not to damage the fish by trying to work the hook out. Cut the line, plop the fish overboard and tie on another hook—preferably a very small sized #4 English style. The hook will rust out and the fluke will live to fight another day.

Watch the weather reports before venturing out after fluke. If the wind is blowing south two days in a row, even on an 80-degree day, the water will be cold (for reasons we don't begin to understand). Cold water drives the fluke out to warmer deeper water and suggests staying home to catch up on chores.

If you do fish for fluke in cold water, follow this example. ML went out with a dozen anglers in the third day of a south blow. In all, fourteen fish were caught on a poor half-day excursion—thirteen of them by Manny. Unlike the other anglers, he kept his reel open at all times, allowing the sinker to slide along bottom in the south-to-north drift but making sure his bait was drifting just off bottom, where the sluggish fluke would be. When he felt even the slightest weight, he slowly let out lots of extra line, then lifted his rod tip slightly to "feel" for fluke (rather than a crab or a sinker stuck in sand). When he felt resistance, he slammed the rod tip high. On thirteen occasions that day, it produced fish.

A 3- to 4-inch tapered strip of squid is the usual bait, but such "white baits" as a length of fluke belly (when legal) or strips of sand shark, sea robin, herring, or mackerel also work well, especially when "sweetened" with a live killy or dead spearing or sand eel.

66 Manasquan Ridge

Directions: Sail due south for 10 miles out of Belmar to the Manasquan Ridge, which is approximately 5 miles off the beach and nearly parallel to the Klondike Banks.

The Manasquan Ridge is a huge underwater structure that in summer is frequented by flotillas of headboats, charter craft, and private boats out for a day of great fishing. The Ridge features several wrecks but no sizable rock piles. However, the wrecks do hold sea bass and blacks and an occasional cod. Fluke also are caught here each fall and mackerel each spring, but the featured fish—indeed, the kings and queens of the ridge—are bluefish, which inhabit the ridge in huge numbers to forage on heavy concentrations of bait.

Figure 23 ■ Sea bass like the one held by Josh Kasper are caught at the Manasquan Ridge in good numbers.
(Photo: Captain Joe Kasper.)

In summer, the blues are usually taken jigging by day and chumming/bait fishing at night. Occasionally bonito and false albacore are hooked, adding to the general excitement.

When going out for blues late each fall, take a quantity of 2- to 6-ounce Ava-style jigs. If cast underhanded, allowed to fall to bottom, and then reeled back to the boat quickly, these jigs will produce most of the bites by blues rather than the undesirable sand sharks ("green-eyed horn dogs"), which swim more slowly and flock to bait.

> Tip: *Most jigs used for bluefish have rubberized teasers fastened onto the hook to simulate little sand eels and to give the lure added motion. We prefer to use a jig with no permanent tail. We fashion our own lures, using 7-inch black/grape-colored jelly worms threaded several inches up the hook. Remember to take a pocket full of worms, since every strike from a blue will mean a worm lost.*

The Manasquan Ridge is especially well remembered for the best catch of blues—in terms of poundage—ever nailed by ML. Before the ten-fish limit was imposed, he boated thirty-six fish ranging in size from 8 to 16 pounds, fishing on the half-day *Norma K III* out of Point Pleasant. The 200 pounds of fresh fillets were distributed to needy families in Somerville, N.J.

67 Hudson Canyon

Directions: The northwest boundary, referred to as "the Tip" of The Canyon, is 73 miles east by southeast of Manasquan Inlet.

The Hudson Canyon is the most famous tuna-fishing grounds off the Jersey coast. Captain Pete Barrett, managing editor of *The Fisherman* magazine, is our favorite authority on this spot. He fishes the canyon from late June until the end of October, primarily for yellowfin tuna and longfin albacore. Anglers also take bigeye tuna and dolphins ("mahi mahi") at The Canyon in large numbers.

The banana-shaped canyon is nearly 20 square miles in size, and runs from 50 to well over 100 fathoms in depth. The southwest corner of The

Canyon is referred to as "The Letters"; the southeast corner is called "The Point."

Captain Barrett and other experienced skippers catch many of their biggest fish while trolling, including the yellowfins and bigeyes, plus white and blue marlin. Pete looks for weed lines, changes in water color, and lobster-pot markers as likely trolling spots. However, the best indicator is a warm-water temperature break, an eddy that is visible only by use of underwater temperature indicators. Trolling is often red hot on one side or the other of such breaks.

Boats troll large plugs or carefully rigged whole ballyhoo—a long silvery fish with a beak. (Breaking off the beak makes the fish move in a more lifelike fashion in the water.) Trolling is best during the daylight hours; the preferred method of fishing the canyon at night is drifting or even anchored at an inside edge of the canyon, using natural bait. (We are reminded that it is illegal to tie onto any of the many lobster-pot marker buoys to hold position.) The bait of choice is whole butterfish, with the same fish used as chum.

> Tip: *Chop the butterfish and hand-feed it overboard so as to lure the fish to your boat, while leaving your whole baitfish as the most attractive offering.*

The Canyon produces huge fish. Captain Barrett's two personal bests were a 314-pound bigeye tuna he took on a trolled green and yellow MoldCraft Hooker plug and a blue marlin with an estimated weight of 450 pounds. The massive blue, which he released unharmed, hit a big trolled "horse" ballyhoo.

Clearly, fishing the canyon requires a big boat with two engines. For huge game-fish excitement, The Canyon is "never-never land." Don't miss it.

79 Barnegat Ridge

Directions: The Ridge is 11 nautical miles out of Barnegat Inlet.

Captain Charles Eble began taking anglers out to The Ridge in 1951 on his boat, *The Doris Mae,* and in the intervening years, he and his sons have fished its massive expanse often and successfully. The Ebles' fares have taken literally tons of sport fish along its miles of irregular shapes, and the skippers have accumulated a storehouse of knowledge about the area. Best of all, their boat is still pounding out of the inlet to this day.

Captain Eble, Jr., notes that the closest section of The Ridge is its northwest corner, 11 nautical miles out, and that it runs nearly parallel to the beach for 2 miles to its southern edge. The depth climbs sharply from 90 feet to 60 feet when you move up onto the ridge, and the fishing action generally climbs just as sharply. Bluefish reign supreme on the ridge, often found in massive numbers. Memorial Day marks the unofficial start of the bluefish season, and blues stay on the ridge through mid-December if water temperatures stay above 50 degrees. Many anglers prefer jigging for bluefish, but Captain Eble recommends chumming with ground mossbunker and baiting with a bunker back or butterfish chunk, night or day.

Tip: If fishing The Ridge on your own boat, carry a capped plastic bottle or two with 80 feet of line tied to at least a pound of lead. When your fish finder picks up fish in numbers, circle to locate the spot and drop a jug or two over as markers.

(In the early 1960s, we saw a charter-boat captain mark a spot by throwing pails of garbage overboard. Our method of using a retrievable plastic bottle is much more environmentally sound.)

Big mackerel also are caught here as they migrate back and forth in spring and fall, as well as bonito and false albacore. In addition, school-sized bluefin tuna and yellowfin tuna are caught on The Ridge, acting as a magnet to boats of all sizes that can negotiate the modest distance from shore. It is important to remember that bluefin tuna are carefully regulated. Know the current regulations before killing one.

If tuna excite more anglers' imaginations, it is still bluefish who provide most of the action. For big blues, don't miss the Barnegat Ridge.

99 Cape May Rips

Directions: Proceed out of Cape May Inlet to the southernmost tip of land (at the mouth of the Delaware). Work the "Rips" south by southeast, from just offshore out to a distance of about 3 miles.

The Cape May "Rips," so called because of odd tidal movement that often appears to be rippling in different directions at the same time, are in fact two spots: the first beginning quite close to shore, the second perhaps 3 miles off of Cape May. The state-record hybrid bass, a 12-pound 7-ounce lunker that may well have come down the Delaware River, was taken on the Rips several years ago by angler David Kmiec. Having battled many 6- and 7-pounders in fresh water, we both look at this record fish with undisguised envy.

The Rips provide outstanding action on bluefish, Boston mackerel, and fluke, especially as the waters start to warm in early spring. Charter and headboats sometimes catch king and Spanish mackerel beginning around April 1, with some of the kings running to double figures in size. This can be exciting fishing.

Once the mackerel run passes further to the north, bluefish show up along the rips in large numbers. Most are in the 1- to 3-pound class, but anglers jigging bucktails or trolling metal often take very large fish.

Fluke also feed on the Rips in good quantity and size, from late April right through the cold weather at the end of fall.

> *Tip: While we generally prefer a plain gold hook with no bangles or beads for this type of fishing, we have found that when drifting squid bait here, your chances are enhanced by adding a small spinner and bead lure just before the eye of the hook. Also, a fillet of sea robin or sand shark often produces good results on a fast drift.*

While fast action on mackerel, fluke, and blues keeps the boats coming back, the real queen of the Rips is the striped bass. The magnitude of the striper fishery in New Jersey marine waters, in numbers of fish and in size, is difficult to exaggerate. In one recent year, the total catch of stripers—including released fish—was estimated at more than 779,000 fish. Since these fish can weigh 50 pounds or more, this fishing is invested with special excitement. (Of note: the largest striper taken in New Jersey waters, a 78-pound 8-ounce beast landed off Atlantic City, is a world record.) Is bass fishing exciting?

The stripers show up at the Rips in late spring, feeding somewhat lethargically in the colder water. The best approach to catching these slow-moving linesides is a bucktail dropped to the bottom and jigged ever so slowly. The fish are simply too lazy to chase a lure whipped up and away from them. But if you lift slowly and bounce your bucktail, you can do extremely well.

Stripers feed more enthusiastically as spring turns to summer. By early fall, they congregate along the Rips in large numbers as they prepare to go up the saltwater rivers to spawn. Live eels are almost irresistible at this season, and a single experienced angler catching and releasing a dozen big fish in a morning is not impossible. When using eels, (1) be sure to keep them in very cold water or sand, so that you can handle them in the hooking process, and (2) make sure to lip-hook them properly (ask the captain or mate if you're not sure) on a striper hook. There are dozens of head-boats and charter boats out of South Jersey ports that fish for stripers daily in the late fall along the Cape May Rips. Any one of them may well deliver a lifetime of memories in a single morning.

Artificial
Reefs

The Division of Fish, Game, and Wildlife's Artificial Reef Program, supported by many private and public contributions (including *The Fisherman* magazine and Clarks Landing Marina) provides extraordinary marine habitat where only barren ocean floor previously existed. Begun with two reefs in 1984, the state's program has expanded to a network of fourteen active ocean reef sites, from Sandy Hook all the way down to Cape May. Some of the reef sites are as large as 4 square miles, and more than 1,000 individual "patch reefs"—measuring a half acre to several acres in size— have been built within these sites. The reefs are constructed of more than 2.2 million cubic yards of dredge rock, concrete, ships and barges, obsolete Army vehicles, and concrete-ballasted tire units. Soon after they are deployed, new reef materials are blanketed by a living carpet of filter-feeding creatures such as mussels and barnacles. Crabs, snails, and shrimp soon follow, and in their turn, tens of thousands of reef fish take up residence on the reefs.

The results are spectacular. In two recent years, recreational anglers caught 1.8 million and 1.1 million fish on reef sites, including bonito, Spanish mackerel, and bluefish as well as all of the more usual reef fish.

Figure 24 ■ Recycling a boat the "new-fashioned" way.
(Photo: N.J. Division of Fish, Game, and Wildlife.)

And the reef fishery continues to grow, with the following six among the most productive.

45 Sandy Hook Artificial Reef

Directions: The reef is located over a 2-mile area running from 2 miles north of the Shrewsbury Rocks down to the rocks, with its closest point to shore being 1 mile from Sea Bright.

The construction of Sandy Hook Artificial Reef was begun in 1985 and is today, by a considerable margin, the largest constructed. It comprises 1,834,000 cubic yards of structure. A significant portion of this reef material was generated as a result of the demolition of the old Palmolive plant in Jersey City. Vast amounts of reinforced concrete and other material generated in the demolition were cleaned, placed on barge after barge, and towed out to sea. (Interestingly, proximity to shore helps determine the types of materials used in reef construction. Sandy Hook and Cape May reefs, being closest to shore of the state's fourteen reef sites, receive mostly construction debris, since deployment is an economically feasible disposal method for contractors.) The material was sunk in predetermined sites in a carefully patterned plan, and virtually overnight, the normal progression began: first microscopic organisms, then tiny shellfish, then barnacles and crabs, then small fish seeking food and shelter, then larger fish to feed on small fish and crabs.

Now the reef is a super site for these fish—especially blackfish, sea bass, and porgies. Captain George Bachart, who fishes the reef regularly, tells us that the blacks are at the site in numbers from spring through late fall, showing a marked preference for green crabs. Sea bass compete for food here mid-spring through mid-fall, and porgies show up in numbers late each summer.

Tip: When fishing for porgies, bring small snelled hooks, size 4 or even 6, and a dozen or so bloodworms. Sandworms work too, as do skimmer clam bellies, but bloodworms are our first choice.

It is important to note that this location is NOT marked by a buoy. If sailing on your own craft, buy a map at a quality tackle store for coordinates, or better yet, hop aboard a headboat making for the Sandy Hook Artificial Reef. You'll be in for an interesting experience.

61 Sea Girt Artificial Reef

Directions: Sail north out of Manasquan Inlet on a heading of 70 degrees for 3.8 miles, or 4.3 miles out of Shark River on compass bearing 158 degrees.

Bill Figley of Fish and Game is generally credited with being the driving force behind the artificial-reef program in the Garden State. The first to be constructed was the Sea Girt Artificial Reef, which was begun more than twenty years ago. One of dozens of early volunteers one memorable day was ML, who—along with other volunteers—lifted, and lifted, and lifted countless airplane-sized tire units weighted with cement construction forms, which were lashed together and plopped overboard. The boat was the *Miss Point Pleasant,* owned by Jack Kennell, proprietor of Ken's Landing and the *Norma K* fleet.

Larger materials were added later, including a half dozen carefully cleansed tugboats and tons of steel and concrete which became available with the destruction of the Route 88 Bridge at Point Pleasant. The result is a dynamite fishery, one that produces fine fishing eight months of the year and only fails to produce when the shallow waters get too cold each winter.

The reef is close enough to shore for small private boats to access easily. The benefit is that anglers can use light tackle, which would just get tangled on a crowded boat. Ling are the main attraction here in the spring as they come in from offshore to feed and spawn. You can easily tell a fish that has just made the long swim; its skin color has a pink tone rather than the usual muddy brown.

Sea bass inhabit the reef in huge numbers although not necessarily huge size. (Remember to check your compendium for size limits before putting one in the cooler.) Porgies and fluke also appear in the fall, and blackfish are caught throughout much of the eight-month season, and, adding to the rich marine diversity of the reef, lobsters are taken here by divers.

Tip: *To rig for blackfish, go with a single whole green crab as bait and two size 2 blackfish hooks tied below your sinker. Snip off the claws and stick one hook in each claw opening, penetrating so that the hooks come out. You won't get a lot of bites on this rig, but when your whole crab is taken, it usually means a hard-fighting black of 5 pounds or more on the other end of your line.*

At least one railroad car has been added to the reef and a number of huge rock piles. The most recently added vessel was a 99-foot Navy tug sunk on the reef by the Fisherman's Conservation Organization in honor of Captain Greg Venturo, who lost his life during a dive.

The artificial reefs are a superb idea that gets better and better with each new deployment of materials by agencies of government working with concerned organizations and private citizens.

78 Garden State North Artificial Reef

Directions: Head on a compass bearing of 172 degrees, 7.7 nautical miles out of Barnegat Inlet.

Garden State North is "one of the best artificial reefs built so far off the Jersey Coast," according to Bill Figley, who knows whereof he speaks. Indeed, Mr. Figley is one of the driving forces behind the entire artificial-reef program.

The Garden State North reef is 6.5 miles offshore and covers more than a full square mile in depths ranging from 66 to 83 feet. Thus far, seventy-nine separate placements of material have taken place—a total of 36,000 cubic yards of structure. The materials include large numbers of concrete-weighted tires along the western half of the reef and fourteen sunken vessels and a number of Army tanks, interspersed with more tire units, throughout its eastern half. Significantly, some of the patch reefs—that is, individual sites within the larger reef sites—have been sponsored by corporations and groups interested in aiding New Jersey Fish and Game in this exceptionally productive program. One example is Penn Reels, which has sponsored the sinking of railroad cars, concrete, Army tanks, and boats at the upper center of the entire reef.

Close by is the John Dobilas Reef site, where employees of the McGraw-Hill company, the McGraw-Hill Foundations, and Charles Chillemi provided financial assistance for the sinking of the 165-foot Navy tanker U.S.S. *John Dobilas.*

The Garden State North Reef is home to all of the typical reef fish, including blackfish, porgies, ling, bergal, and sea bass. The best drift fishing on this reef is found along its entire left side. Often this fishing is superb.

Tip: When embarking on a trip to the reef, make sure to take along some crabs for bait, especially if you are after blackfish. While blacks also like squid and clams, these baits can be more easily cleaned off your hooks by marauding bergals, which swarm around the structure at bottom.

83 Garden State South Artificial Reef

Directions: Run a compass bearing of 198 degrees, 11.1 miles out of Barnegat Inlet. Even closer, sail 64 degrees for 9.1 miles out of Little Egg.

Garden State South is thus far one of the smaller artificial reefs in the Division of Fish, Game, and Wildlife's reef program. It comprises about 10,000 cubic yards of sunken structure. Nonetheless it holds plenty of fish and is a favorite destination for public and private boats alike.

The reef, which covers .6 mile in depths of 57 to 63 feet, is made up largely of weighted tire units, sunken boats, and discarded Army vehicles. The latter include five tanks contributed through a joint military operation that provides obsolete combat vehicles, which are cleaned, trucked to the water, then loaded onto barges for their final ocean voyage. The tanks are completely colonized by mussels, barnacles, and hydroids, often in less than two years on the sea floor, and provide excellent reef habitat for many species of fish and shellfish. At least eight vessels also have been sunk on the reef.

Figure 25 ■ An Army tank on its way to becoming part of an artificial reef.
(Photo: N.J. Division of Fish, Game, and Wildlife.)

The portion of this reef that currently is best for drift fishing is its mid-left area. Sea bass are the primary target here. Averaging about 1 to 3 pounds, they fight extremely well for their size.

> *Tip: Fish for these scrappy bass with a stiff rod and at least 20-pound test monofilament line, since they will head for the nearest bottom structure upon being hooked and must be turned quickly after the strike.*

Fluke and other bottom-feeding species, including tautog and porgies, also are caught here in great numbers.

We recommend against diving along this section. The bottom is considerably more open here and there is less for the diver to see. In addition, anglers drifting the area will be greatly inconvenienced by your marker buoy.

91 Atlantic City Artificial Reef

Directions: Follow a 142 degree heading out of Absecon Inlet for 8.8 miles.

Fifty-five to ninety-four feet below the wave-tossed surface, sections of the Atlantic City Reef might resemble a military vehicle parking lot, since obsolete Sheridan and M-60 tanks comprise many of the hard substrate materials deployed here. Cleaned, barged to the site, and sunk by units of the New Jersey National Guard and the U.S. Naval Reserve, these vehicles have quickly become covered with marine life and now provide excellent habitat for a wide variety of reef fish.

Measuring more than 4 square miles in area, this reef also is the site of a number of sunken boats plus concrete-ballasted tire units—all attractive to marine life. At this writing, more than 52,000 cubic yards of structure have been placed at strategic points in this large reef site.

Like the other artificial reefs, the Atlantic City Reef is loaded with all of the usual reef fish, including black sea bass, ling, porgy, tautog, and bergals. In addition, fine populations of "tropical water" fish are found here in late summer and early fall. Recently, ML discovered this to his considerable pleasure when, fishing with Captain Applegate out of Atlantic City, he and a friend caught hard-fighting amberjack and triggerfish along with many of the more usual denizens of the New Jersey reefs.

Tip: While squid is the usually prescribed bait, we find that fresh or even frozen surf ("skimmer") clam works a good deal better for just about all of these fish.

100 Cape May Artificial Reef

Directions: Follow a 128 degree heading out of Cape May Inlet for 9.1 nautical miles.

The Cape May Artificial Reef is a superb fishery for a wide variety of bottom fish and one of the finest examples of the Division of Fish, Game, and Wildlife's artificial reef construction program.

The reef is 3 miles long and almost a mile wide, situated in an average of 60 feet of water. It is a 9-mile run straight out of Cape May—one of fourteen artificial reefs that have been strategically located within easy boat range of New Jersey's ocean inlets. Captain Neil Robbins, who often makes for the reef out of Sea Isle City, told us that components of the Cape May reef range from sixteen obsolete ships, cleaned by the government and scuttled at the site, to massive volumes of cleansed cement and steel made available when the old Benjamin Franklin Bridge Roadway was demolished. In addition, thousands of large tires weighted with cement were lashed together—largely by Southern State Correctional Facility inmates—and added to the fish-friendly structure.

Large numbers of crabs, shrimp, snails, and other organisms that provide forage for fish have taken up residence in the tires and crevices of the sunken ships. In their turn, tautog, ling, porgies, sea bass, and other adult fish have been attracted, providing wonderful fishing for boats anchored along the reefs. Boats also troll the reef, especially where large shipwrecks attract bonito, some Spanish mackerel, and bluefish.

Party boats and private boats alike make the Cape May reef their destination for wonderful fishing. The primary quarry is sea bass and thousands of anglers go home with nice catches, from April through December. Squid is the most effective bait, but bass also are taken on skimmer clams. Early and late in that eight-month period, blackfish (tautog or "tog") are found on the reef, showing a marked preference for crab. Either a piece of green crab or an entire fiddler or China back crab works well on size 2 or 4 blackfish hooks.

> *Tip: In early spring, tog have soft mouths; consequently softer bait works better than crab. The belly of a clam will produce fish, as will 2-inch pieces of bloodworm or sandworm.*

Triggerfish, characteristically found in warmer waters to the south, show up on the reef in summer and are caught in numbers. These fish are excellent table fare and especially easy to skin-peel fillet. Porgies, small amberjack, and fluke cruise the reef as the water cools. Expect especially good catches of fluke—many 3 pounds or heavier—when drifting the tire portion of the reef. Try longer leaders, size 3/0 English bend hooks and long pieces of squid single-hooked (rather than bunched, as for sea bass) when fluke are your target.

As noted, in the most recently surveyed year, recreational anglers caught an estimated 1.1 million fish on artificial reefs. It is thus little wonder that so many of us applaud Fish and Game (and cooperating anglers and other private citizens) for a wonderful and highly productive program.

Index

Numbers in **boldface** refer to key map, pp. xx–xxi

About the Authors

Manny Luftglass and Ron Bern have been friends and fishing partners for well over thirty years, during which time they have spent thousands of hours together competing for the most fish and the largest in the best fishing spots in New Jersey. Separately or together, they have fished well over half of the 100 locations covered in this book—some of them hundreds of times. They have carefully researched the remaining spots—talking with countless anglers and guides, poring over maps and reviewing libraries of information from the state and from books and outdoor publications.

Manny is the author of ten successful books on fishing. Alone he authored *Gone Fishin' in Spruce Run Reservoir, Gone Fishin' in Round Valley Reservoir, Gone Fishin' in N.J. Saltwater Rivers and Bays,* and *Gone Fishin' for Carp.* With friend Joe Perrone, Jr., he wrote *Gone Fishin' with Kids.* He writes feature articles for the weekly angling publication *The Fisherman,* and was formerly the mayor of Somerville, New Jersey.

Ron is the author of four published books, including a novel, *The Legacy,* and three nonfiction books on business subjects. He recently has completed two new novels and is at work on a third. He is a career writer and was for many years president of a management consulting firm in New York. He is a member of "Who's Who in America."

Look for more books from Manny and Ron.